Modern Intelligent Instruments -Theory and Application

Authored by
Changjian Deng
Chengdu University of Information Technology
Chengdu, Sichuan
China

Modern Intelligent Instruments -Theory and Application

Author: Changjian Deng

ISBN (Print): 978-981-14-6024-1

Published by Bentham Science Publishers Pte. Ltd. Singapore. All Rights Reserved.

First published in 2020.

need for a court order if at any point you breach any terms of this License Agreement. In no event will any delay or failure by Bentham Science Publishers in enforcing your compliance with this License Agreement constitute a waiver of any of its rights.

3. You acknowledge that you have read this License Agreement, and agree to be bound by its terms and conditions. To the extent that any other terms and conditions presented on any website of Bentham Science Publishers conflict with, or are inconsistent with, the terms and conditions set out in this License Agreement, you acknowledge that the terms and conditions set out in this License Agreement shall prevail.

Bentham Science Publishers Pte. Ltd.
80 Robinson Road #02-00
Singapore 068898
Singapore
Email: subscriptions@benthamscience.net

BENTHAM SCIENCE

CONTENTS

FOREWORD

This book introduces the main principles and typical applications of modern intelligent instruments.

This book introduces the main principles of modern intelligent instruments, including circuit signal and system, uncertainty and data analysis, signal processing, measurement and human-computer interaction, and the other features of this book, including MATLAB, PROTEUS, ARDUINO, *etc*. This book introduces the application of intelligent instruments in somatosensory networks, robots, and equipment status.

It is suitable for senior Undergraduates' textbooks and reference books for scientific and technological personnel.

Fucun Qu
Chengdu University of Information Technology
Chengdu, Sichuan
China

PREFACE

Measurement and Intelligent instruments play an important role in ordinary life. The industrial development promotes the technology innovation of measurement and intelligent instruments, both in theory and application. For example, the advance in the theory of measurement includes the zero definition and datum quantity deformation research in fields of classical definition; additive conjoint measurement, empirical relation, the rule of mapping, and the representation condition of measurement research in the fields of representational theory; the unambiguous meaning of the measurement problem research in quantum mechanics and the advance in the theory of intelligent instrument include audio, video, speech, image, communication, geophysical, sonar, radar, medical, and musical signals research in their application research; cyber-physical systems, the Internet of things, cloud computing, cognitive computing researched in the fields of intelligent instrument design, and so on. Meanwhile, the theory and application of intelligent instruments promote the developments of industry 4.0 or industry 2025.

The "Modern Intelligent Instruments: theory and application" has ten chapters, Chapter 1 is "The concept of intelligent instrument", which mainly discusses the property, scale, and level of measurement, the concept of intelligent instrument; Chapter 2 is "signal detection and analysis technology of intelligent instrument", which mainly discusses the noise analysis technology that is applied in data acquired system circuit design, analog-digital convert device design and medical, weak signal detection, and meanwhile introduces the modern filter technology and weak signal detection technology; Chapter 3 is "the data processing technology of intelligent instrument"; in this chapter, the measurement uncertainty and the data processing algorithms used in industrial intelligent instrument are introduced, meanwhile the inverse problem and intelligent computing are discussed; Chapter 4 is "the performance analysis of intelligent instrument"; in this chapter, accuracy theory, reaction speed consideration, especially the reliability engineering of intelligent instrument are discussed, mainly introduces reliability assessment, reliability program plan, reliability design and reliability analysis; Chapter 5 is "the data communication and display technology of intelligent instrument", which mainly discusses the IIC, SPI, and HMI technology of intelligent instrument; Chapter 6 is "Proteus and its simulation examples"; in this chapter, proteus is used in hardware and software simulation of intelligent instrument design; Chapter 7 is "Arduino and MATLAB in intelligent instrument", which mainly discusses the advance of Arduino and MATLAB used in intelligent instrument; Chapter 8 is "The application of intelligent instrument: the mobile intelligent instrument", which mainly discusses the advance of the wearable and IOT intelligent instrument,; Chapter 9 is "The advance of the intelligent instrument applied in on-line equipment monitoring system", for example, the fault tolerance, Fault analysis, fault identify problem, and so on; and Chapter 10 is "The application of intelligent instrument: mobile robot instrument", which mainly discusses the moving robot (for example the smart small car and micro UAV).

CONFLICT OF INTEREST

The authors confirm that the content of this book has no conflict of interest.

Changjian Deng
Chengdu University of Information Technology
Chengdu, Sichuan
China

ACKNOWLEDGEMENT

This book is supported by the students of Measurement & Control Technology and Instrument Major of the Chengdu University of Information Technology. It is also supported by the teachers of the school of Control Engineering, the administrative departments, and the Party and government offices in the College. We also thank Professor Fu Lin of the Chengdu University of Industry for his support and guidance.

## CHAPTER 1

The Concept of Intelligent Instrument

Abstract: One of the most fundamental principles in science and technology is that the discovery can be reproduced or the results can be measured. So, at the beginning of the book "modern intelligent instruments-theory and application", the introduction of measurement, the intelligent instrument and its composition, and the example of an intelligent instrument are present.

Keywords: Intelligent instrument, Measurement, Metrology, Oscilloscopes.

1.1. THE INTRODUCTION OF MEASUREMENT

1.1.1. The History of Instruments

1) The major early measuring and measuring instruments.

Weighers and timers are the earliest measuring instruments of human beings. They reflect the early understanding and living needs of human beings. Evidence of the use of the balance in 2500 B.C. has been found, and the earliest indication of the use of the balance in ordinary trade was 1350 B.C. From 300 B.C. to 100 B.C., the magnetic compass, or orientation instrument, was invented.

2) From the Middle Ages to 1500, there were some precision instruments in the world. At this time, astronomical instruments were relatively accurate, for example, the equatorial theodolite, goniometer, leveling instrument, and so on. In 780, Muslim Mint workers placed the balance in an airtight container. By comparing the weighing results two times, the resolution of balance was almost 1/3 mg after countless swings. This is the ancestor of an analytical balance.

3) In the late 15th century, with the development of natural science, early scientific instruments were gradually formed in different backgrounds and forms, mainly optical instruments, thermometers, pendulum clocks, and mathematical instruments.

4) Since the mid-20th century, with the development of automatic control theory and technology, digital instruments based on A/D conversion have been dev-

eloped rapidly. Meanwhile, with the development and maturity of computers, communications, software, new materials, and new technologies, artificial intelligence and on-line measurement and control have become possible, which makes the instrument intelligent, virtualized, and networked. Digital Instruments, Intelligent Instruments, Personal Computer Instruments, Virtual Instruments, and Network Instruments represent the mainstream of modern scientific instruments in the 20th century.

1.1.2. The Main Concepts of This Subsection

Measurement is to describe the observed phenomena with quantity according to a certain law, that is, to make a quantitative description of things. Measurement is the process of quantifying non-quantified objects. In mechanical engineering, measurement refers to the experimental cognitive process of comparing the measured and standard quantities with measurement units in numerical value, to determine the ratio of the two [1].

Measurement is a set of operations to determine the measurement to be measured. With the help of special equipment, the measured results are directly or indirectly compared with the known units of the same kind, and the measured results expressed by both numerical values and units are obtained. Quantity is the attribute that distinguishes things qualitatively and quantitatively. In other words, measurement is practiced to obtain information about the object to be tested.

There are four elements in the measurement process: 1) measuring objects, 2) measuring principle, 3) the suitable measuring instruments, 4) and measured results. Measurement is a process obtained by the experiment and can reasonably be given a quantity of one or more measures to the measuring objects. The measured properties reflect the different properties of objects from different sides, and they provide objective possibilities for measurement. Or, measurement is the use of quantity to describe observed phenomena and quantify things under certain laws and criteria [2].

Measuring standards and measurements are clearly defined, which are the two basic conditions of measurement. Metrology is an activity to achieve unity of units and accurate and reliable measurement values [3].

At present, legal units of measurement based on the international unit system are widely used in the world. International units include basic units, export units and auxiliary units. The original SI units for the seven basic physical quantities were Table **1-1** [3].

Table 1-1. Seven basic physical quantities of SI units.

Quantity	Unit	Symbol
time	second	s
length	metre	m
mass	kilogram	kg
electric current	ampere	A
temperature	kelvin	K
amount of substance	mole	mol
luminous intensity	candela	cd

There are some advances in the concept of measurement which are as follows: Representational theory, Information theory and Quantum mechanics.

Representational theory: Representation theory makes an algebraic object table a more specific matrix and makes the operations in the original structure correspond to matrix operations, such as matrix synthesis, addition and so on. The beauty of representation theory is that it can transform abstract algebraic problems into linear algebraic operations. Representation theory is also applied in natural science. The problem of symmetry cannot be separated from the group, and the study of the group depends on its representation [4].

Information theory: Information theory is a science that studies the law of measurement, transmission and transformation of information employing mathematical statistics. It mainly studies the common law of information transmission in communication and control systems and the basic theory of optimum solution to the problems of information limitation, measurement, transformation, storage and transmission. Information is the increase of certainty - inverse Shannon information definition; Information is the indication of matter, energy and information - the inverse of Wiener's definition of information; Information is a collection of identifications of things and their attributes [5].

Quantum mechanics: Quantum theory is one of the two cornerstones of modern physics. Quantum theory provides a new way of observing, thinking and expressing nature. Quantum theory reveals the basic laws of the microcosmic material world and lays a theoretical foundation for atomic physics, solid physics, nuclear physics, particle physics and modern information technology. It can well explain the atomic structure, the regularity of the atomic spectrum, the properties of chemical elements, the absorption and radiation of light, the infinite separability of particles and information, *etc.* Especially its openness and uncertainty inspire more discoveries and creations [6].

1.1.3. Development of Measurement and Metrology

In this subsection, some advances in the National Institute of Metrology are discussed [7].

1) The second. Time is defined by atomic oscillation frequency. Therefore, frequency stability and frequency accuracy become an important concept of time measurement. For example, time stability measurement is using the time interval counter between the measured clock and the reference clock (*e.g.* seconds), respectively; frequency stability measurement is using two oscillators with different but similar frequencies.

2) The meter: the length measurement is mainly based on the laser which can radiate stable wavelength. The wavelength reproduction accuracy of the He-Ne laser with methane absorption and frequency stabilization is the highest. Its wavelength is 3.39 mm. In practical application, it is usually transmitted by two kinds of physical standards: working wavelength standard, and line segment standard.

3) The kilogram: On November 16, 2018, the 26th International Congress of Measurement (CGPM) adopted Resolution 1 on the Revision of the International System of Units (SI) by a vote of all member states, including China. According to the resolution, the definitions of four SI basic units, namely, kilogram, ampere, Kelvin and Moore, will be changed to constant definitions, which will come into effect on May 20, 2019. One kilogram will be defined as "the unit of mass corresponding to the Planck constant of $6.62607015 \times 10^{-34}$ J.s". The principle is to convert the mechanical force required to move a mass of 1 kilogram into electromagnetic force expressed by Planck constant and then calculate the mass by mass-energy conversion formula.

4) The ampere: scientifically, the amount of electricity passing through any cross-section of a conductor in a unit time is called current intensity, which is referred to as current. The current symbol is I; its unit is ampere (A). On November 16, 2018, the 26th International Conference on Metrology adopted the resolution of "Revising the International Unit System". It defined 1 ampere as "the current intensity generated by the movement of charges in one second $(1/1.602176634) \times 10^{19}$".

5) The kelvin: Temperature is a physical quantity that represents the degree of heat and cold of an object, and microscopically it is the degree of intense thermal motion of an object molecule. Temperature can only be measured indirectly by some properties of the object that vary with temperature. On November 16, 2018,

the International Conference on Metrology adopted a resolution that Kelvin was defined as "the thermodynamic temperature corresponding to the Boltzmann constant of $1.380649*10^{-23}$J.K^{-1}". The new standard definition came into force on May 20, 2019.

6) The mole: A unit of quantity of matter. On November 16, 2018, the International Conference on Metrology adopted a resolution that 1 mole would be defined as "the amount of substance in a system that contains exactly 6.02214076 x 10^{23} basic units such as atoms or molecules". At the same time, the Avogadro constant is modified to 6.02214076 x 10^{23}.

7) Candela: Candela is one of the seven basic units of the International System of Units (SI). Abbreviated as "Kan", symbol cd. It is the intensity of a light source in a given direction. The light source emits monochromatic radiation with a frequency of 540 x 10 Hz, and the radiation intensity in this direction is 1/683 Watt/sphericity.

8) Siemens Process Instrumentation, Keithley, Agilent develop many measurement solutions and products [8 - 10].

1.2. THE INTELLIGENT INSTRUMENT AND ITS COMPOSITION

1.2.1. Eight Kinds of Test & Metering Instruments

Instruments are the general name of various instruments and devices used to observe, monitor, measure, verify, record, transmit, transform, display, analyze, process and control material entities and their properties. And the meter is mostly the measurement device applied in the industrial field [11, 12]

Class by eight kinds of test & metering instruments:

1. Geometric quantity: length, angle, morphology, mutual position, displacement, distance measuring instruments, *etc.*
2. Mechanical Quantity: All kinds of force measuring instrument, hardness meter, acceleration and speed measuring instrument, torque measuring instrument, vibration measuring instrument, *etc.*
3. Thermal capacity: temperature, humidity, flow measurement instruments, *etc.*
4. Optical parameters: such as photometric meter, spectrometer, colorimeter, laser parameter measuring instrument, optical transfer function measuring instrument and so on.
5. Ionizing radiation: various radioactivity, radionuclide metering, X, gamma rays and neutron metering instruments, *etc.*

6. Time frequency: A variety of timing instruments and clocks, cesium atomic clocks, time frequency measurement instruments and so on.
7. Electromagnetic Volume: AC, DC ammeter, voltmeter, power meter, RLC measuring instrument, electrometer, magnetic parameter measuring instrument, *etc.*
8. Electronic Parameters: oscilloscope, signal generator, phase measuring instrument, Spectrum analyzer, Dynamic Signal Analyzer and other radio parameter measuring instruments.

Category by Professional categories: Industrial automation instrumentation and control system Scientific instruments Electronic & amp; electrical measuring instruments; Medical devices; *etc.*

Instrumentation engineers are responsible for integrating the sensors with the recorders, transmitters, displays or control systems, and producing the instrumentation diagram. They may be responsible for calibration, testing and maintenance of the system [12].

1.2.2. Learning Example: Oscilloscopes

The oscilloscope is composed of four parts: vertical controls, trigger, horizon controls, and display. Now, the LCD panel is used for display. The normal oscilloscopes display (CRT) has focus, intensity, beam finder controls.

One can use the button or switch of vertical to select the different amplitudes of the displayed signal.

One can use the button or switch of the horizon to select the different time base (Seconds-per-Division)of the displayed signal.

One can use the button or switch of the trigger to select the different start events of the sweep of the displayed signal [8].

Application notices: there is a small portable instrument often used, for example, the PC-based oscilloscopes(AD2).(www.digilent.com.cn/)

1.2.3. The Intelligent Instrument

As shown in Fig. (1-1), the intelligent instrument comes from the combination of computer technology and measuring technology, it is a measurement (or detection) instrument containing micro-computer or microprocessor, it has the function of storage, operation, logic judgment and automatic operation of data, and has a certain intelligent function (manifested as the extension or enhancement of intelligence, *etc.*) [12].

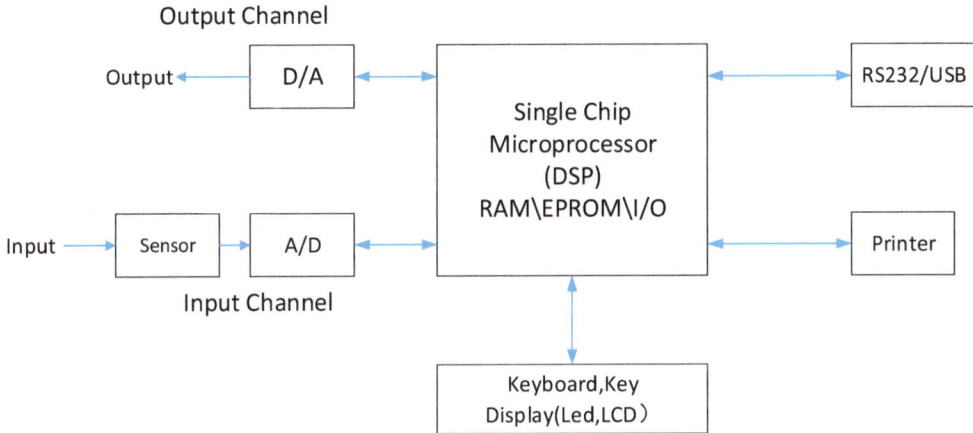

Fig. (1-1). The diagram of embedded intelligent instrument.

The intelligent instrument has the functions of perception, memory, analysis, identification, thinking, and behavior similar to human or some higher animals, its perceptual parts are detection system, memory parts are storage system, analytical thinking parts are the logic operation and control system of the computer, and the behavior component is the executing mechanism of the instrument.

It includes an embedded intelligent instrument and platform intelligent instrument.

The embedded intelligent instrument combines a single-chip or multi-chip microcomputer chip into an instrument [13].

And the platform intelligent instrument is the application of an extended measuring instrument with a personal computer (PC) as the core. It includes Personal Computer Instruments (PCI), microcomputer card instruments, and so on.

Features of Intelligent instruments include [13]:

1) Software control of the measurement process:

CPU→ Software Control Measurement process;

"Hard to Soft" → flexible, strong reliability.

2) Data processing:

Random error, system error, nonlinear calibration, and other processing →
improve measurement accuracy ;

Signal analysis of digital filtering, correlation, convolution, deconvolution,
amplitude spectrum, phase spectrum, power spectrum, *etc.* → Provide more high-
quality information.

3) Versatility: One machine multi-use such as an intelligent power demand
analyzer.

1.3. THE EXAMPLE OF INTELLIGENT INSTRUMENT

Fig. (**1-2**) is the problems in intelligence instrumentation [14].

Fig. (**2a**) is the perception of an intelligence instrumentation design, the initial is
the problem statement. It is an evolutionary design and becomes a single well-
defined problem.

Fig. (**2b**) shown: the design includes the problem statement, data interpretation
and decision making.

Fig. (**2c**) shows the expert system in instrumental systems design.

Fig. (**2d**) shows a variety of instruments and interpreting results based on one or
more analyses.

1) Problem Statements: Ideally a problem statement is analogous to the standard
hypothesis in statistical analysis.

2) Knowledge Base: to operate as an intelligent system, it must have the
knowledge base necessary to solve each of its problem statements.

3) Expert Systems Driven Multivariate Data Systems:

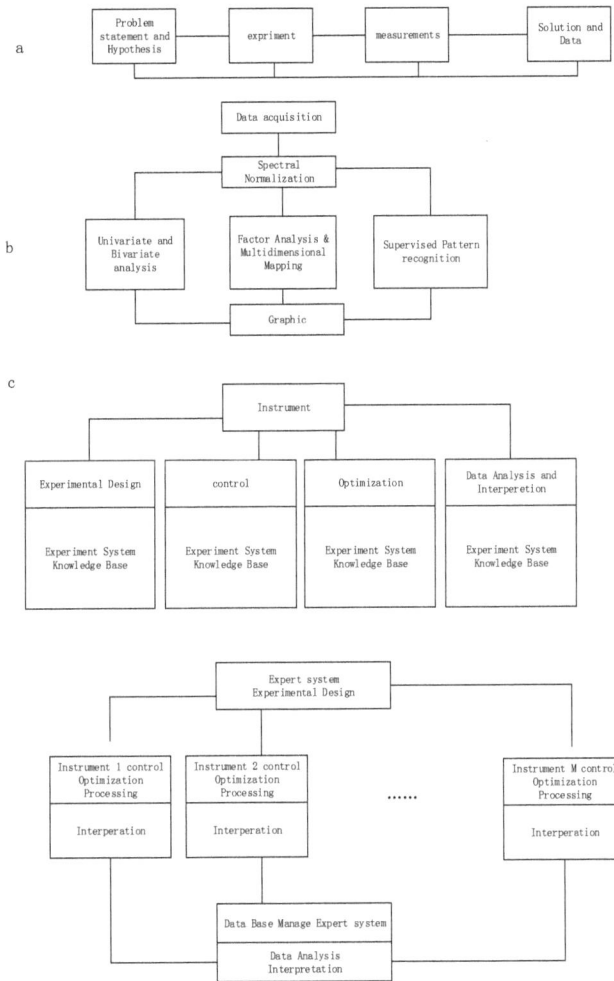

Fig. (1-2). The example of an intelligent instrument design problem.

PROBLEM

1-1 Searching for three kinds of intelligent sensors on the internet, and briefly describing their principles.

1-2 Brief description of the realization method of intelligent sensor

1-3 Searching for an example of an intelligent instrument design problem

REFERENCES

[1]   Measurement, *available from https://en.wikipedia.org/wiki/Measurement.*[Accessed: 2019.2.1].

[2]   L. Fu, and C.J. Deng, *Theory, Design and Application of Intelligent Instruments.* Southwest Jiao tong

University Press: China, Chengdu, 2014.

[3]   R. Köhler, "The International Vocabulary of Metrology, 3rd Edition: Basic and General Concepts and Associated Terms", In: *Why? How? Transverse Disciplines in Metrology.* ISTE, 2010.

[4]   R. Luce, "Representational Measurement Theory", In: *Stevens' Handbook of Experimental Psychology,* 2002.

[5]   Thomas M Cover, and J. A. Thomas, *Elements of information theory,* 2003.

[6]   M.A. Nielsen, "Characterizing mixing and measurement in quantum mechanics", *Phys. Rev. A,* vol. 63, no. 2, pp. 184-184, 2001.
        [http://dx.doi.org/10.1103/PhysRevA.63.022114]

[7]   S. Schreppler, N. Spethmann, N. Brahms, T. Botter, M. Barrios, and D.M. Stamper-Kurn, "Quantum metrology. Optically measuring force near the standard quantum limit", *Science,* vol. 344, no. 6191, pp. 1486-1489, 2014.
        [http://dx.doi.org/10.1126/science.1249850] [PMID: 24970079]

[8]   https://new.siemens.com/global/en/products/automation.htmlAccessed: 2019.2.1

[9]   https://www.tek.com/Accessed: 2019.2.1

[10]   https://www.agilent.com/Accessed: 2019.2.1

[11]   Instrumentation, *available from https://en.wikipedia.org/wiki/Instrumentation.*Accessed: 2019.2.1

[12]   Instrument, *available from https://en.wikipedia.org/wiki/Instrument.*Accessed: 2019.2.1

[13]   D.F. Chen, and Q. Lin, "The Intelligent Instrument, China, Beijing: The China Machine Press.2014. A. M. Harper, S. A. Liebman. "Intelligent Instrumentation", *J. Res. Natl. Bur. Stand.,* vol. 90, no. 6, pp. 453-464, 1985.

CHAPTER 2

The Signal Detection and Analysis Technology in Intelligent Instrument

Abstract: The fundamental modern intelligent instrument is signal detection and analysis technology. Although it is the classic content of intelligent instruments, the noise analysis technology, the weak signal detection technology developed largely in many test fields. Therefore, in Chapter 2, the structure principle of data acquired system, the noise analysis technology, and the weak signal detection technology are introduced.

Keywords: Analog to digital converter, Boxcar integrator, Data acquired system, Lock-in amplifier, Multiplexer, Noise, Structure principle, Sample and hold, Simulation, The adaptive filter.

2.1. THE STRUCTURE PRINCIPLE OF DATA ACQUIRED SYSTEM

2.1.1. Multiplexer

Analog electronic switches include transistor switches, photoelectric coupling switches, junction field-effect transistor switches, and insulated gate field effect switches.

The principle of an electronic multiplexer is shown in Fig. (2-1). It is a switch with multiple-input and single-output. The schematic of the multiplexer is in the left of the figure, the equivalent switch is present on the right [1].

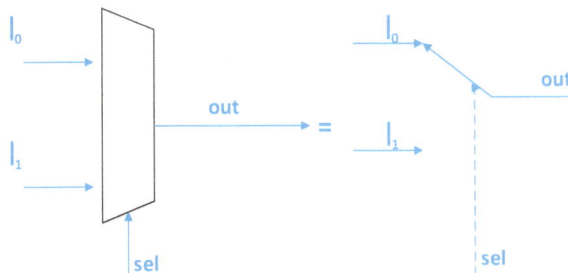

Fig. (2-1). Schematic of a 2-to-1 Multiplexer.

The Multiplexer can be built by transmissions gates as shown in Fig. (**2-2**). From the symmetrical structure, the different level of control signal 'sel' determines whether the input signal 'a' pass to output 'q' or the input signal 'b' pass to output 'q' [1].

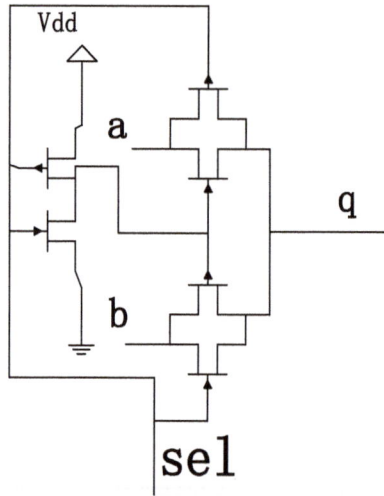

Fig. (2-2). CMOS Schematic of a 2-to-1 Multiplexer.

2.1.2. Sample and Hold

Sample Holder refers to the circuit in "sampling" or "holding" state under the control of the input logic level. The output of the sampling state circuit tracks the input analog signal, and the output of the holding state circuit maintains the instantaneous input analog signal at the end of the previous sampling until it enters the next sampling state [2].

The sample holder consists of an analog switch, a memory element (holding capacitor) and a buffer amplifier. The structure of S/H is shown in Fig. (**2-3**).

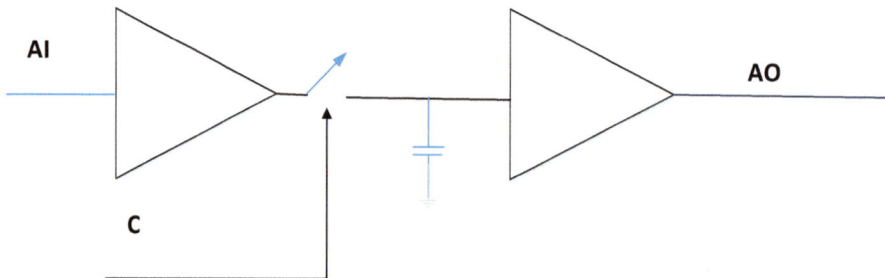

Fig. (2-3). The typical S/H circuit.

There are three types of Sample Holder: series type, feedback type and capacitance correction type. In the series type of S/H, a Voltage Follower is added in front of AI, whose output misalignment error is the algebraic sum of two Operational Amplifier misalignment errors; the feedback type of S/H adds the feedback loop based on the series S/H; and the capacitance correction type of S/H improvs the circuit to solve the accuracy problem in the holding stage.

2.1.3. An Analog-to-digital Converter

A/D converter is a device that converts the analog voltage or current signal into digital quantity. It is an interface between the analog system and the digital system(or PC) [3]. A/D converter means it converts sampled analog signals into discrete code, there are three processes: sampling, quantizing and coding.

The main performance index of an ADC includes resolution, quantizing error, offset error, full-scale error, linearity, absolute accuracy, and conversion rate. The convert rate determines its bandwidth, its sampling rate. The SNR of an ADC is influenced by the resolution, linearity and accuracy, aliasing and jitter.

Nyquist–Shannon sampling theorem: A signal can be reconstructed only if its sampling rate is greater than twice the bandwidth of the signal.

1) The successive approximation analog-to-digital converter

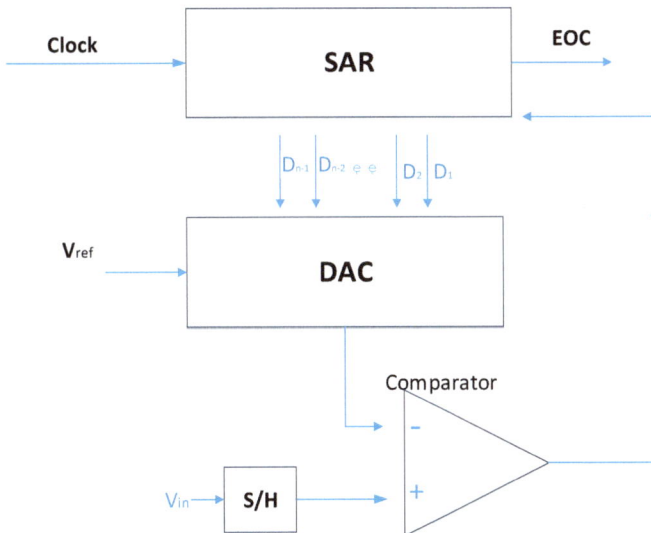

Fig. (2-4). The block diagram of Successive approximation ADC.

The structure of successive approximation analog-to-digital converter is shown in Fig. (2-4).

The successive approximation A/D converter compares the input analog voltage with different reference voltage several times. When comparing the input digital quantity of DAC, it successively determines the "0" and "1" states of each digit, so that, the converted digital quantity can approximate the corresponding value of the input analog quantity numerically [4]. There are four-part in the successive approximation A/D converter, the first is the successive approximation register(SAR), the second is the comparator, the third is control signal(the signal of start ST, the signal of the end of the conversion EOC), and the fourth is DAC.

Example: Conversion process of the four bits successive approximation A/D converter.

A. Start Conversion: Approximation Register Output Zero-out, 4-bit DAC before the start of conversion, Output V0 = 0. The conversion control signal VL = 1 starts.

B. successive approximation register output 1000: Under the action of CLK's first clock pulse, the highest bit output of the successive approximation register is 1, and the rest bit output is 0, that is, the output of the successive approximation register is 1000.

C. Under the action of the second clock pulse, the sub-high position 1 is sent to the comparator to compare with the input signal Vi, to determine whether the sub-high position 1 should be retained or not. Compare them one by one until the lowest comparison is completed and the conversion is completed.

If the reference voltage is 5, the binary weights assigned to each bit, starting with the MSB, are 2.5, 1.25, 0.625, 0.3125.

2) The dual slope integrating ADC

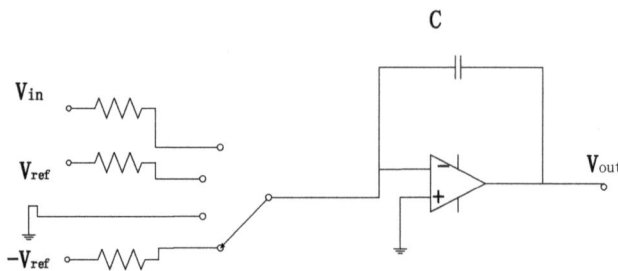

Fig. (2-5). Run-up dual-slope integrating ADC.

The main structure of the dual slope integrating ADC is shown in Fig. (**2-5**) [5]. Its principle is described in the following:

A. Initial phase: START = 0, the control signal of the control logic output makes the counter clear 0 (the overflow bit of the counter is cleared at the same time), while the control logic control analog switch S0 closes so that the capacitor C can discharge fully.

B. The first stage of integration: Integration start: START = 1, control logic output control signal (state combination of S1 and S2) control analog switch S and Vin to turn on, so that the integrator reverse integration of Vin. Integrator starts reverse integration (first integration): If Vin > 0, there is Vout < 0, Vc > 0, S and Vin connected at the same time control logic control counter start counting (counting pulse cycle is T0) when the counter is full, its overflow bit changes to 1, control circuit according to the state of Vc and overflow bit control analog switch S1 and - V_{REF} connected, the same. The counter starts counting from zero.

C. Integrator starts forward integration (second integration): When V0 rises to slightly greater than 0, Vc becomes a low level, which makes control logic output control signal and control counter stop counting. At this point the counter count value is the A/D conversion value. Because there are two integral processes, called double integral A/D conversion.

If the interference of double integral A/D is superimposed on the input signal during the integral period, because the interference is generally symmetrical, the output of the integrator will take its average value to play the role of filtering, improve the anti-interference ability, and have a wide range of practical applications. But because the conversion accuracy depends on integration time, the conversion speed is slow.

The normal converting time of double integral A/D is the times of 50 or 60Hz, to decrease the power interfering noise.

2) The high speed analog-to-digital converter

Parallel comparative A/D converter: The output of each flip-flop is directly fed to the input of the priority encoder. According to the function of the priority encoder, only the high-level output of the highest-level comparator is coded. So, the corresponding output code d2d1d0 of the encoder can be obtained. This is the digital quantity corresponding to the analog quantity.

2.1.4. The Simulation of ADC and Example

In Fig. (**2-6**), the analog signal outputs the digital signal through the component of the delay, switch, ADC conversation, buffer, nonlinearity, and quantizer, situation. This one interleaved high-speed ADC model can show the processing of its conversation and interleave.

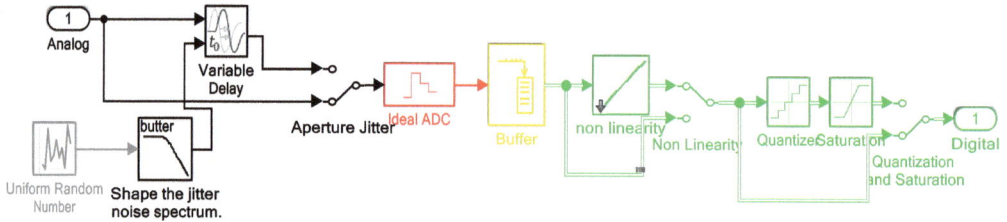

Fig. (2-6). One interleaved high-speed ADC.

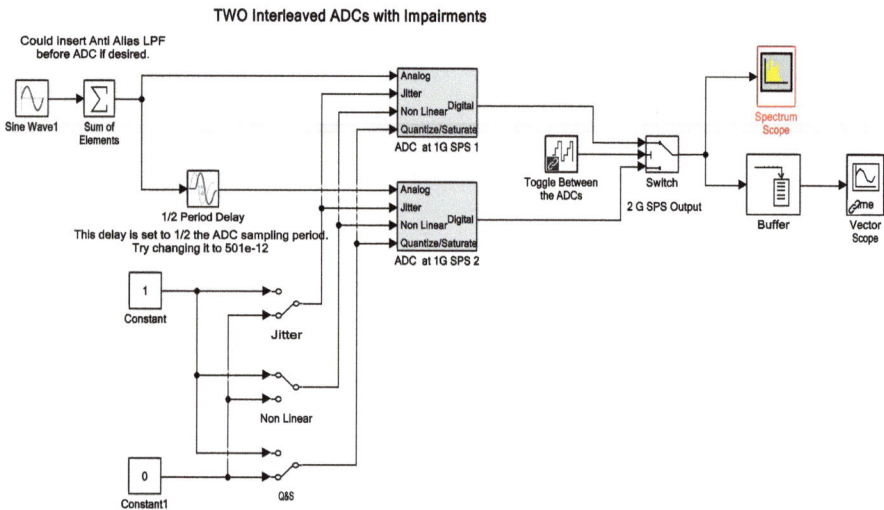

Fig. (2-7). The Simulink model of two interleaved ADCs with impairments.

The Simulink model of two interleaved high-speed ADC is shown in Fig. (**2-7**), This Simulink model of two interleaved high-speed ADC model is based on one interleaved high-speed ADC.

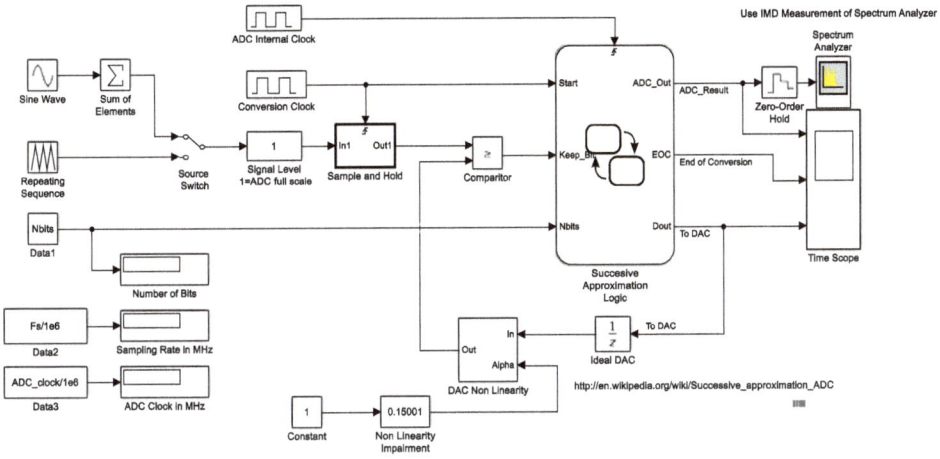

Fig. (2-8). The Simulink model for successive approximation analog to digital converter.

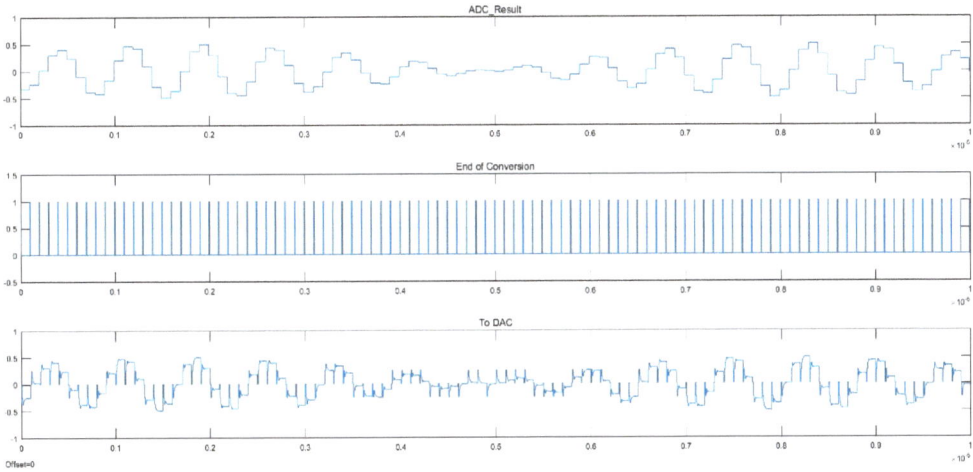

Fig. (2-9). The ADC result, end of conversion, to DAC signals of SAR ADC.

The Simulink model of successive approximation ADC is shown in Fig. (2-8). The ADC results, the end of conversion, to the DAC signal is shown in Fig. (2-9).

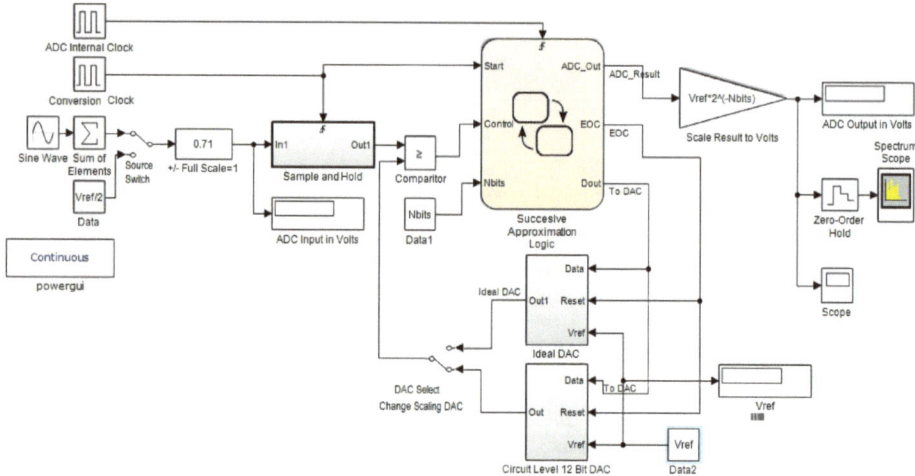

Fig. (2-10). The Simulink model of 12bits successive approximation analog to digital converter.

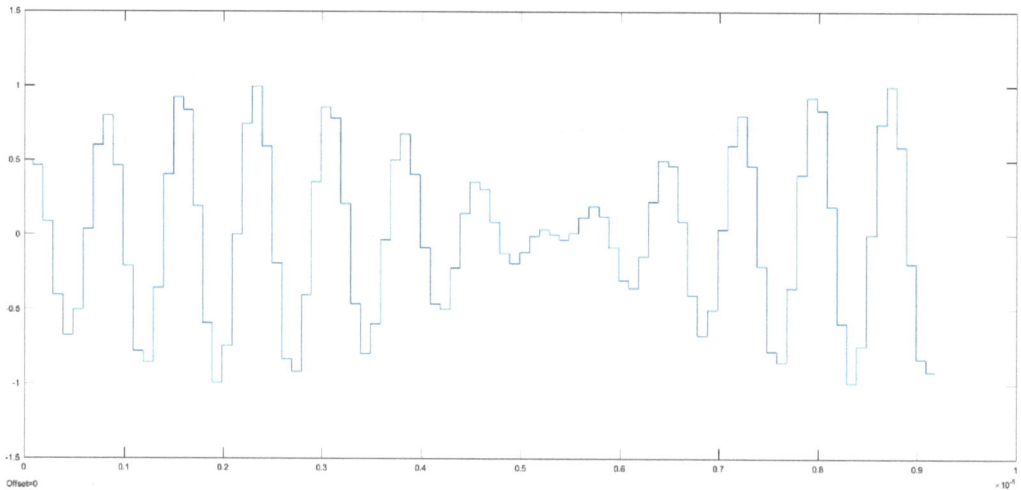

Fig. (2-11). The time scope of the Successive Approximation Analog to Digital Converter(12 bits).

The Simulink model of 12bits successive approximation analog to digital converter is shown in Fig. (**2-10**). The time domain waveform of 12bits successive approximation analog to digital converter is shown in Fig. (**2-11**).

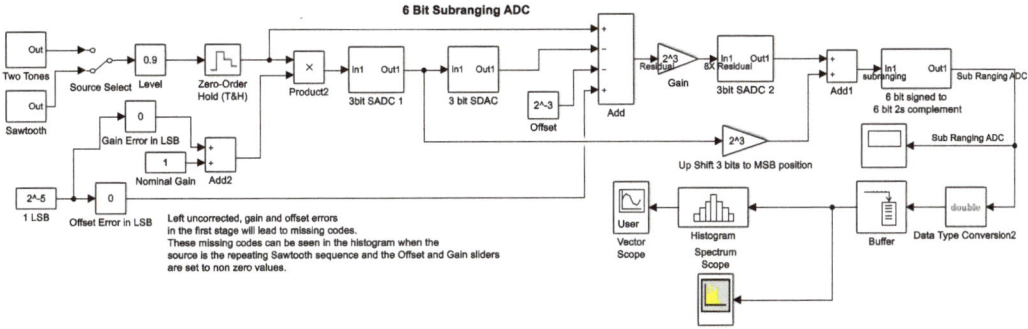

Fig. (2-12). The 6 Bit Sub ranging ADC.

The Simulink model of the 6 Bit Sub ranging ADC is shown in Fig. (**2-12**).

2.2. NOISE ANALYSIS TECHNOLOGY

The word "noise" means any "unwanted sound" [7].

The concept of Gauss white noise: White refers to a noise signal whose power spectrum is constant; probability p(x) is a Gauss function when the amplitude of Gauss refers to various values. In the field of communication, additive white Gauss noise refers to a noise signal whose spectrum components obey uniform distribution (*i.e.* white noise) and whose amplitude obeys Gauss distribution. It is named for its additivity; its amplitude obeys the Gauss distribution and is a kind of white noise. The noise signal is an ideal noise signal for easy analysis. The actual noise signal can be approximated only by the characteristics of Gauss white noise in a certain frequency band. Because the AWGN signal is easy to analyze and approximate, in the field of signal processing, the noise performance of signal processing system (such as filter, low-noise high-frequency amplifier, wireless signal transmission, *etc.*) can be simply analyzed (such as signal-to-noise ratio analysis). Generally, it can be assumed that the noise generated by the system or the noise signal disturbed by the system at a certain frequency. Gauss white noise under segment or constraint conditions. Additive white Gauss noise is only one kind of white noise, and Poisson white noise, *etc* [7, 8]:

2.2.1. Noise Types and Color Noise

1) Johnson–Nyquist noise (sometimes thermal, Johnson or Nyquist noise): it is the electronic noise caused by thermal stirring when the charge carriers (usually electrons) in conductors reach equilibrium state, independent of the applied

voltage. Generally, the noise is deduced by statistical physics, which is called the wave dissipation theorem. In this paper, the medium is characterized by generalized impedance or generalized polarizability. The thermal noise of an ideal resistor is close to white noise, that is, the power spectral density is almost uninterrupted throughout the spectrum (but not at very high frequencies). When the bandwidth is limited, the thermal noise approximates the Gauss distribution [9].

2) Shot noise is the noise caused by the non-uniformity of electron emission inactive devices (such as vacuum tubes) of communication equipment. Also known as shotgun noise. Particle noise is caused by the dispersion of the carriers that form the current. In most semiconductor devices, it is the main source of noise. At low and intermediate frequencies, shot noise is independent of frequency (white noise). At high frequencies, the spectrum of shot noise becomes frequency dependent. The current mean square of shot noise is proportional to electron charge, total DC current and bandwidth [9].

3) Flicker noise, also known as 1/f noise, slow random fluctuations of emitted electrons are caused by local fluctuations of devices (*e.g.* local inhomogeneity on the surface of photocathodes). These changes usually occur at lower frequencies (the upper limit of frequencies is about 500 Hz). This kind of noise is called flicker noise. At low frequencies, the scintillation noise is several times larger than the shot noise, but it can be reduced by increasing the working frequency band, avoiding the low frequency band and carefully fabricating optoelectronic devices [9].

4) Burst noise is an electronic noise that occurs in semiconductor and ultrathin gate oxide film. It is also known as Random Telegraph Noise (RTN), Popcorn Noise, Pulse Noise, Bistable Noise or Random Telegraph Signal (RTS) Noise. Popcorn noise occurs at low frequencies (usually f < 1kHz). We already know that heavy metal atom pollution is the cause of popcorn noise. In failure analysis, experts usually examine devices with more burst noise carefully. Failure analysis will look for minor defects that can cause sudden noise [9].

5) Transit-time noise: This kind of noise is caused by the problem of the transmission medium in the process of signal transmission, such as bad contact of connectors, bad material of signal lines, crosstalk of ground current and so on. Among them, ground current crosstalk is often neglected. Because most of the civil audio equipment adopts unbalanced transmission mode, the outer shield layer of the signal line participates in the signal transmission. Usually, the shield layer is connected with the "ground" of the audio equipment. The ground of most audio equipment relates to the outer shell of the equipment and relates to the

"ground" provided by the power supply line of the residence [9].

6) Coupling noise: This kind of noise is caused by coupling. To realize the transmission of energy and signal, the method of connecting each functional circuit is a coupling circuit. Generally, coupling circuits usually have one or more functions, such as filtering, energy storage, isolation, impedance transformation, *etc*. Coupling refers to the close coordination and interaction between input and output of two or more circuit elements or circuit networks, and the transmission of energy from one side to the other through interaction [9].

7) Pink noise: pink noise is the most common noise in nature. In short, the frequency component power of pink noise is mainly distributed in the middle and low frequency bands. From the waveform point of view, pink noise is fractal, and audio data has the same or similar energy in a certain range. From the power (energy) point of view, the energy of pink noise decreases from low frequency to high frequency, with a curve of 1/f, usually decreasing by 3 decibels per 8 degrees. Pink noise is the most commonly used sound for acoustic testing. Pink noise can be used to simulate sounds such as waterfalls or rain [9].

8) red noise: also known as brown noise, is caused by Brownian motion, also known as random moving noise. Thus, the term Brown correctly refers to Robert Brown, who discovered Brownian motion, rather than brown from color; the term red noise is analogous to white noise because of its high energy intensity at the red end of the visible spectrum. The energy spectral density of Brownian noise is inversely proportional to its frequency, which means that the noise has greater energy at lower frequencies, even more than pink noise. When the frequency increases 8 times, the energy of Brown noise decreases by 6 decibels (that is, 20 decibels for every order of magnitude increase in frequency) [9].

9) Blue noise is also called azure noise. In the limited frequency range, the power density increases by 3 dB per frequency doubling with the increase of frequency (density is proportional to frequency). For high frequency signal, it belongs to benign noise [9].

10) Orange noise. This kind of noise is quasi-static. In the whole continuous spectrum range, the power spectrum is limited and the number of zero power narrowband signals is limited. These zero-power narrowband signals are concentrated in the center of the note frequency of any correlated note system. By eliminating all consonants, these remaining spectra are called "orange" notes [9].

11) Grey noise: The noise is similar to the psychoacoustical equivalent loudness curve (such as the reverse A-weighted curve) in the given frequency range, so the noise level is the same at all frequency points [9].

12) black noise: Black noise (static noise) includes the output signal of an active noise control system after eliminating an existing noise. In the limited frequency range above 20 kHz, the noise with constant power density is like the white noise of ultrasound to a certain extent. This kind of black noise, like "black light", is too frequent to be perceived, but it still has an impact on you and your surroundings. It has f-beta spectrum with beta > 2. According to experience, the noise is harmful. In signal processing, we often refer to Dirac function or unit pulse, which refers to a signal with zero width and infinite high level. However, pulses with infinite low level and infinite high level cannot be found, but signals with bandwidth and power density can be generated according to different requirements, and then these signals are superimposed on the test object so that we can observe which part of the signal is absorbed or which part of the signal will produce resonance [9].

2.2.2. Noise Simulation Model or Implement Codes

The following Matlab code of noise simulation model are retrieved from [10], the Figs. (2-13-2-15) are the running results after little change.

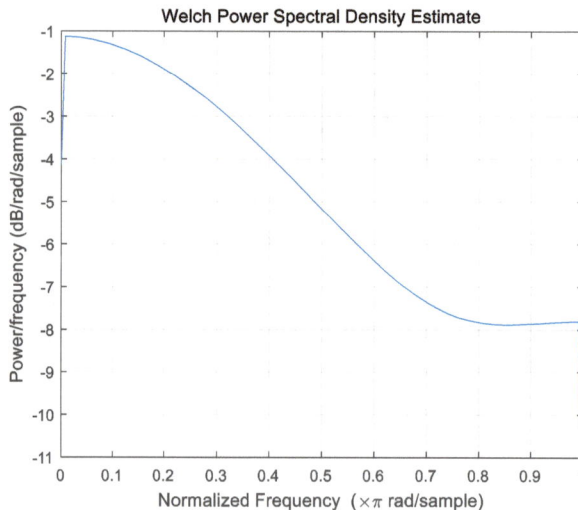

Fig. (2-13). The welch power spectral density of the red noise signal.

The MATLAB code for red noise:

……

x = randn(1, M);

```
% FFT

X = fft(x);

% prepare a vector with frequency indexes

NumUniquePts = M/2 + 1; % number of the unique fft points

k = 1:NumUniquePts;

X = X(1:NumUniquePts);

X = X./k;

% prepare the right half of the spectrum - a conjugate copy of the left

% one except the DC component and the Nyquist component - they are unique,

% and reconstruct the whole spectrum

X = [X conj(X(end-1:-1:2))];

% IFFT

y= real(ifft(X));

% ensure that the length of y is N

y = y(1, 1:N);

% form the noise matrix and ensure unity standard

% deviation and zero mean value (columnwise)

y = reshape(y, [m, n]);

y = bsxfun(@minus, y, mean(y));

y = bsxfun(@rdivide, y, std(y));

……

y=pinknoise(1,32);
```

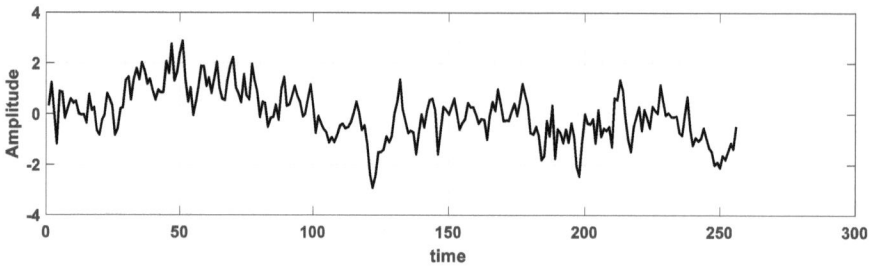

Fig. (2-14). The pink noise signal.

The MATLAB code for pink noise:

……..

```
% generate white noise sequence
x = randn(1, M);
% FFT
X = fft(x);
% prepare a vector with frequency indexes
NumUniquePts = M/2 + 1; % number of the unique fft points
k = 1:NumUniquePts; % vector with frequency indexes
% manipulate the left half of the spectrum so the PSD
% is proportional to the frequency by a factor of 1/f,
% i.e. the amplitudes are proportional to 1/sqrt(f)
X = X(1:NumUniquePts);
X = X./sqrt(k);
% prepare the right half of the spectrum - a conjugate copy of the left
% one except the DC component and the Nyquist component - they are unique,
% and reconstruct the whole spectrum
X = [X conj(X(end-1:-1:2))];
```

% IFFT

y = real(ifft(X));

% ensure that the length of y is N

y = y(1, 1:N);

% form the noise matrix and ensure unity standard

% deviation and zero mean value (columnwise)

y = reshape(y, [m, n]);

y = bsxfun(@minus, y, mean(y));

y = bsxfun(@rdivide, y, std(y));

......

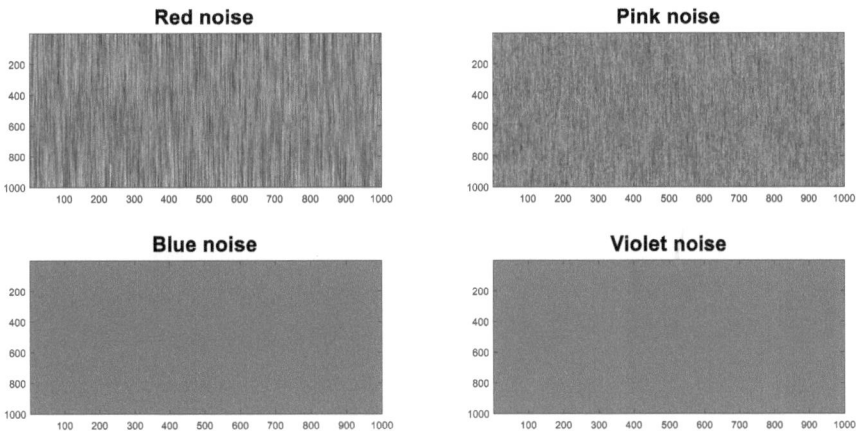

Fig. (2-15). The color noise image.

2.2.3. Noise Analysis Technologies

1) Noise in circuit design

The diagram design for the data acquired system.

A. Why preprocessing circuits in signal conditioning is a low-noise amplifier circuit.

Fig. (2-16). The reason for preamplifier based on noise analysis.

In Fig. (**2-16**), the input noise VIN for the main circuit is obtained by its output noise VON.

$$V_{IN} = V_{ON}/K \qquad (2\text{-}1)$$

Then, the output noise VON' for the system circuit is obtained by the input noise of the main circuit VIN and the input noise of the preamplifier VIN0.

$$V'_{ON} = \sqrt[2]{(V_{IN0}K_0K)^2 + (V_{IN}K)^2} \qquad (2\text{-}2)$$

In formula (2-2), the K0 is amplified ratio of the preamplifier and K is amplified ratio of the main circuit. The total input noise of system VIN' is (2-3)

$$V'_{IN} = \frac{V'_{ON}}{K_0K} = \sqrt{(V_{IN0})^2 + (V_{IN}/K_0)^2} \qquad (2\text{-}3)$$

If we want the VIN'<VIN,in order to VIN'<VIS, then we get (2-4).

$$V_{IN} > \sqrt{(V_{IN0})^2 + (V_{IN}/K_0)^2} \qquad (2\text{-}4)$$

And

$$V_{IN0} < V_{IN}\sqrt{1 - (1/K_0)^2} \qquad (2\text{-}5)$$

Then the preamplifier should fit 1) K0>1; 2) and its noise level is lower than (2-5) [10].

This process is more simplified, there are some notes:

Normally, the IC manufactory defines the input noise and output noise differently. An analysis of this subsection, the input noise more like the signal, it can be an amplifier, and the lots of intrinsic noise of the device cannot be an amplifier, unlike (2-5), the output signal noise ratio were better if there is a preamplifier in its frontend. The output noise is (2-6), the output signal is (2-7), the signal-noise ratio is (2-8). So obviously, the ideal lower noise amplifier can improve the S/N of the system. More, the real case is the worst in both (2-5) and (2-8), for, in amplifier IC device, the input signal must be large (or larger) than all kinds of noise, then it should be amplified and cannot be contaminated.

$$V'_{ON} = \sqrt{(V_{IN0})^2 + (V_{IN})^2} \qquad\qquad \textbf{(2-6)}$$

$$V_{OS} = K_0 \cdot V_{IS} \qquad\qquad \textbf{(2-7)}$$

$$\left(\frac{S}{N}\right)' = \frac{K_0}{\sqrt{1+\left(\frac{V_{IN0}}{V_{IN}}\right)^2}} \cdot \left(\frac{S}{N}\right) \qquad\qquad \textbf{(2-8)}$$

B. The sequence of filer or preamplifier circuit of preprocessing circuit in signal conditioning

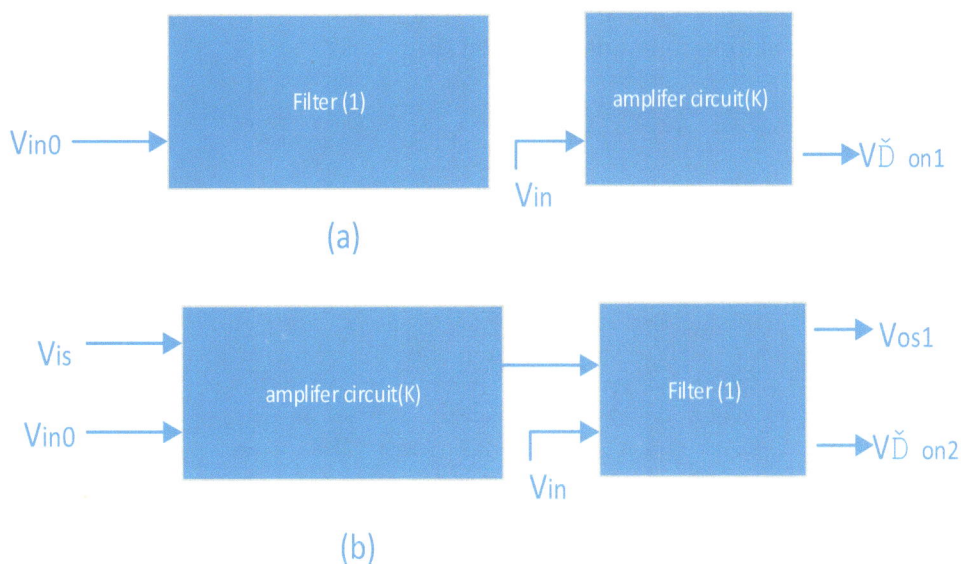

Fig. (2-17). The two different circuit structure in the signal conditioning circuit.

The problem of B is: in Fig. (**2-17**), which is better, (a) or (b).

In (a):

$$V'_{IN1} = \frac{\sqrt{(V_{IN}K)^2+(V_{IN0}K)^2}}{K} = \sqrt{(V_{IN})^2 + (V_{IN0})^2} \tag{2-9}$$

And in (b):

$$V'_{IN2} = \frac{\sqrt{(V_{IN})^2+(V_{IN0}K)^2}}{K} = \sqrt{(\frac{V_{IN}}{K})^2 + (V_{IN0})^2} \tag{2-10}$$

For K>1, then VIN2'<VIN1', the amplifier being the front circuit is better than the filter being the front circuit [11].

C. Noise calculation

Noise calculation often uses the noise spectral density (nV/√Hz), the method is a calculation of the root sum of squares or computing the RMS noise of each noise source separately, but there are some notes should notice. Fig. (2-18) shows the NSD and RTO calculation methods of noise [12].

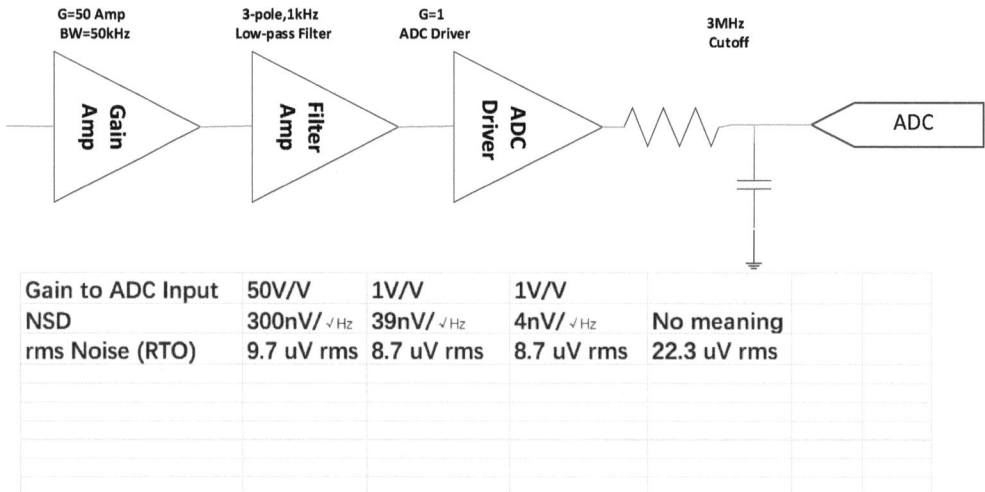

	G=50 Amp BW=50kHz	3-pole,1kHz Low-pass Filter	G=1 ADC Driver	3MHz Cutoff	
Gain to ADC Input	50V/V	1V/V	1V/V		
NSD	300nV/√Hz	39nV/√Hz	4nV/√Hz	No meaning	
rms Noise (RTO)	9.7 uV rms	8.7 uV rms	8.7 uV rms	22.3 uV rms	

Fig. (2-18). Justification for using RMS noise rather than spectral density for noise calculations.

2) Noise in device design

A. Sigma-delta ADC

A sigma-delta ADC consists of analog low-pass filter, analog-Delta modulator and a digital decimation filter. The analog signal is transformed into a band-limited analog signal by an analog low-pass filter. Then, the analog sigma-delta modulator quantizes the band-limited analog signal into a low-resolution digital signal separated from the quantized noise spectrum at a sampling frequency much higher than the Nyquist frequency of the signal band and then filters it with a digital low-pass filter. Quantization noise besides signal frequency band and the sampling frequency is reduced to Nyquist frequency to obtain high resolution digital signal.

Main features of a delta-sigma ADC: oversampling, noise shaping and decimation filter [13].

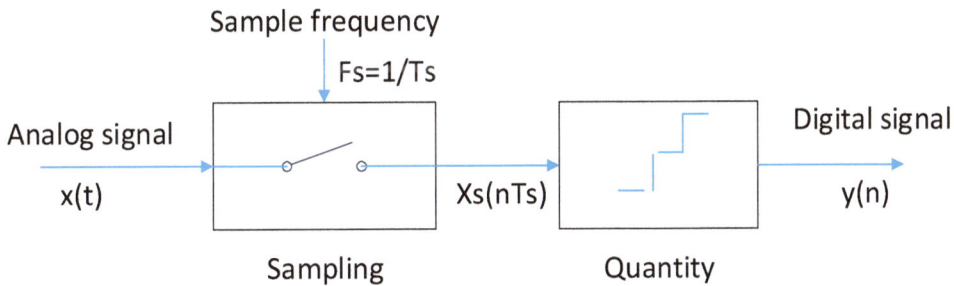

Fig. (2-19). Sampling processing.

In A/D sampling processing, the analog signal $x_s(t)$ is calculated by (2-20).

$$x_S(t) = \sum_{n=-\infty}^{\infty} x(t)\delta(t - nT_S) \qquad (2\text{-}20)$$

In the frequency domain, it is described as (2-21).

$$X_S(\omega) = 1/T_S \sum_{k=-\infty}^{\infty} X(\omega - k\omega_S) \qquad (2\text{-}21)$$

The square of quantization noise is (2-22),q is 1LSB.

$$\sigma_e^2 = \frac{1}{q} \int_{-q/2}^{q/2} e^2 de = \frac{q^2}{12} \qquad (2\text{-}22)$$

Its noise spectral density is (2-23).

$$N(f) = \frac{q^2}{12F_S} \qquad (2\text{-}23)$$

And in the baseband frequency fb, the noise spectral density is (2-24)

$$N_B = \int_{-f_B}^{f_B} N(f)df = \frac{2f_B}{F_s} \cdot \frac{q^2}{12}$$

(2-24)

It is shown in A,B of Fig. **(2-20)** [12].

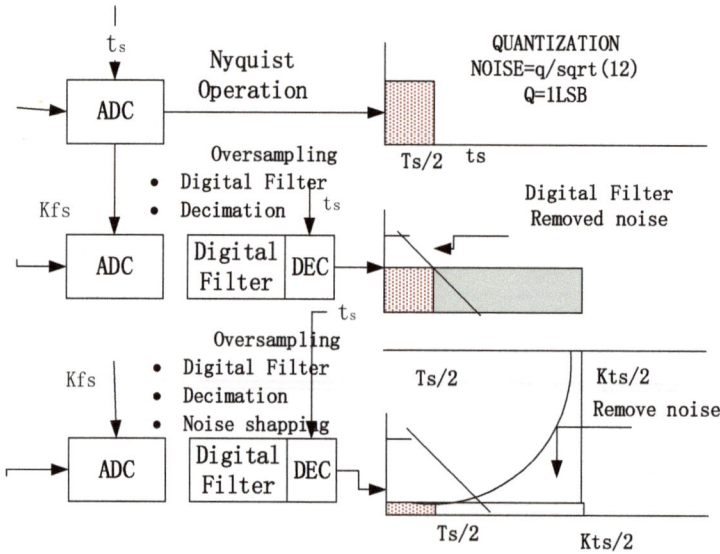

Fig. (2-20). The process of oversampling, digital filtering, noise shaping, and decimation.

The principle of noise shaping is described as Fig. **(2-21)** [13]:

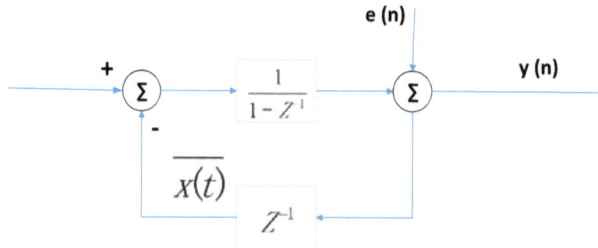

Fig. (2-21). The first order of Z domain delta-sigma modulation.

The output is

$$Y(Z)=X(Z)+(1-Z^{-1})E(Z)$$

(2-25)

The input is:

$$x(t) - \bar{x}(t) \tag{2-26}$$

The encoding of output is shown in Fig. (**2-22**), and the simple reason for noise shaping in Fig. (**2-23**).

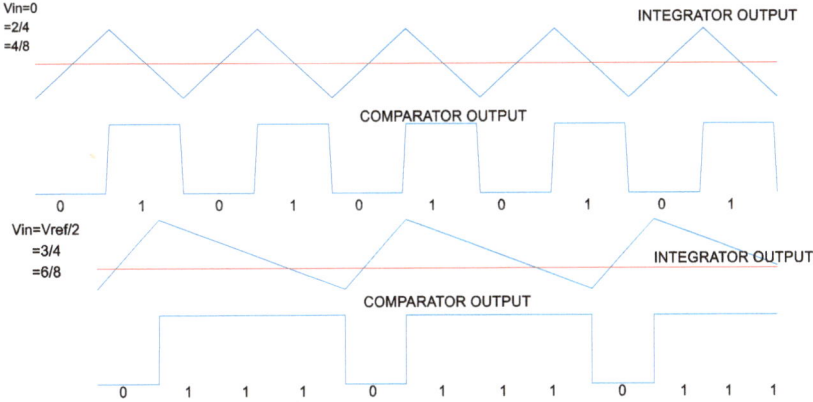

Fig. (2-22). The sigma-delta modulator waveforms.

The Simulink model is shown in Fig. (**2-23**).

(a) The first order sigma-delta modulator

(b) The second order sigma-delta modulator

(a) The first order sigma-delta modulator

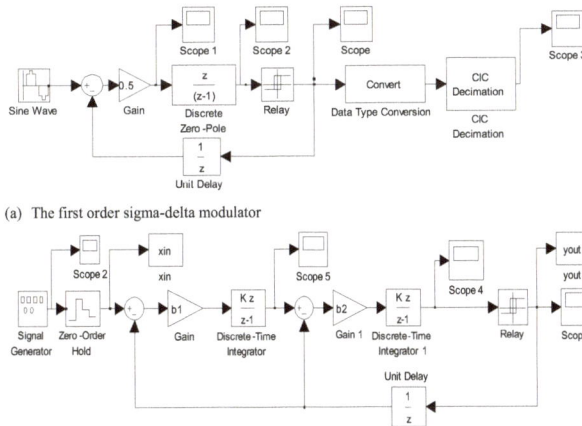

Fig. (2-23). The Simulink model in a sigma-delta modulator.

B. Sampling Oscilloscope

To realize very high-speed sampling, a sampling oscilloscope is used. It acquires and stores the samples, and uses these samples to rebuild the waveform, it can operate the frequency up to 50G Hz. There are two kinds of sampling methods: real time sampling and equivalent time sampling. Equivalent-time sampling includes random and sequential sampling. Using store and rebuild technology, the equivalent time sample can display a very high frequency signal. For example, to display the repetitive natural phenomenon or man-made events, it captures a little bit of information in each repetition, then rebuilds the signal using digital processing technology, so, the frequency of measuring signal can be much higher than the oscilloscope's sample rate [14].

The structure of the sampling oscilloscope is different from the normal oscilloscope, in a sampling oscilloscope the sampling bridge is in the front of the sampling circuit, and in conventional oscilloscope the sampling bridge is in the next of amplifier and attenuator. This is shown in Fig. (**2-24**) [14].

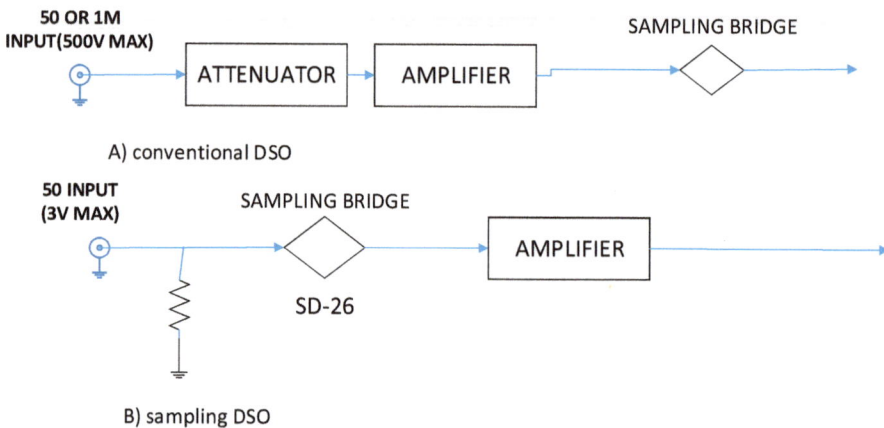

Fig. (**2-24**). The basic structure of scope.

2.3. THE WEAK SIGNAL DETECTION

1) A lock-in amplifier

Lock-in amplifier is a voltmeter that can measure the amplitude and phase of signals buried in noise. General amplifiers amplify the signal and amplify the noise at the same time, so they cannot detect the signal buried in the noise, while lock-in amplifiers can detect the amplified signal and suppress the noise at the same time [9].

Phase-locked amplifier (also known as phase detector) is an amplifier that can separate specific carrier frequency signals from highly disturbing environments (signal-to-noise ratio can be as low as - 60 dB, or even lower). Lock-in amplifier was invented by Robert H. Dick, a physicist at Princeton University. Phase-locked amplifier (PLA) technology came out in the 1930s and entered the stage of commercial application in the mid-20th century. This electronic instrument can extract signal amplitude and phase information in a very strong noise environment. The phase-locked amplifier uses a homodyne detection method and low-pass filtering technology to measure the amplitude and phase of the signal relative to the periodic reference signal. The phase-locked measurement method can extract the signal in the specified frequency band centered on the reference frequency and effectively filter out all other frequency components. Today, the best phase-locked amplifiers on the market have a dynamic reserve of up to 120 dB, which means that these amplifiers can achieve accurate measurements when the noise amplitude exceeds a million times the expected signal amplitude. For decades, with the continuous development of science and technology, researchers have developed many different application methods for PLA. Nowadays, PLA is mainly used as a precise AC voltage meter and AC phase meter, noise measurement unit, impedance spectrometer, network analyzer, spectrum analyzer and phase discriminator in PLL. The related research fields cover almost all wavelength ranges and temperature conditions, such as coronal observations under all-solar conditions, fractional quantum Hall effect measurements, or direct imaging of interatomic bonding properties in molecules. The functions of PLA are extremely rich and varied. Like a spectrum analyzer and oscilloscope, PLA is indispensable and has become a core tool in various laboratory equipment, such as physics, engineering and life sciences. The principle of PLA is shown in Fig. (2-25).

The PLA works according to the orthogonality principle of the sinusoidal function. Specifically, when a sinusoidal function with one frequency 'f' multiplies with a sinusoidal function with another frequency 'v' and then integrates the product (the integration time is much longer than the period of the two functions), the result is zero. If the two functions are equal and the two functions are in the same phase, the average value is equal to half of the product of the amplitude.

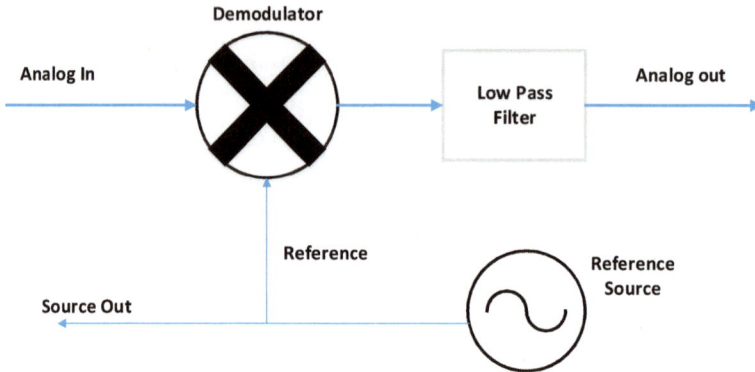

Fig. (2-25). The basic block diagram of a Lock-In Amplifier.

In essence, a lock-in amplifier takes the input signal, multiplies it by the reference signal, and integrates it over a specified time, usually on the order of milliseconds to a few seconds. The resulting signal is a DC signal, where the contribution from any signal that is not at the same frequency as the reference signal is attenuated close to zero.

A. Self-correlation functions

$$R_{xx}(\tau) = \lim_{T \to \infty} \frac{1}{T} \int_{-\frac{T}{2}}^{\frac{T}{2}} f(t)f(t - \tau)dt \qquad (2\text{-}27)$$

Self-correlation functions: It represents the correlation between the random signal $f(t)$ and the same signal after the time interval τ is delayed. When the stochastic function does not contain periodic components, Rxxτ is the largest at τ=0 and monotonically decreases with the increase of τ, τ→∞, Rxxτ approaches the square of the average of $f(t)$, and if the average of $f(t)$ is 0, Rxxτ approaches 0 with the increase of τ. Since the average value of Gaussian noise is 0, the self-correlation function of noise is 0.

B. The correlation functions

$$R_{xy}(\tau) = \lim_{T \to \infty} \frac{1}{T} \int_{-\frac{T}{2}}^{\frac{T}{2}} f_1(t)f_2(t - \tau)dt \qquad (2\text{-}28)$$

Correlation functions: It represents the correlation between the random signal $f(t)$ and the other different signals after the time interval τ is delayed. And the two different random variables (such as signal and noise) are independent of each other. The cross-correlation function will be a constant if an average is 0 (such as noise), the correlation function is 0 everywhere, for it is equal to the product of the average of two random functions.

C. Lock in amplifier

In the field of weak signal detection, the signal amplitude is small (10^{-12}-10^{-9}V). The signal input of the phase sensitive detection PSD has an input signal X (t), which is measured by the characteristic signal $x(t) = A_i \sin(\omega t + \emptyset_i)$ and the noise signal ni(t). It is equal to Equation [9].

$$x(t) = A_i \sin(\omega t + \emptyset_i) + n_i(t) \tag{2-29}$$

The reference signal is $r(t) = A_i \sin(\omega t + \emptyset_r)$. .

The PSD is:

$$u_0(t) = \lim \frac{1}{nT} \int_0^{nT} S_i(t) S_r(t) dt$$
$$= \lim \left\{ \frac{1}{nT} \left[\int_0^{nT} A_i A_r \sin(\omega t + \phi_i) \sin(\omega t + \phi_r) dt + \int_0^{nT} n_i A_r \sin(\omega t + \phi_r) dt \right] \right\} \tag{2-30}$$
$$= \frac{1}{2} A_i A_r \cos(\phi_i - \phi_r) = \frac{1}{2} A_i A_r \cos\phi$$

So, one can tune the value of Øi-Ør,and to obtain the max output of PSD:

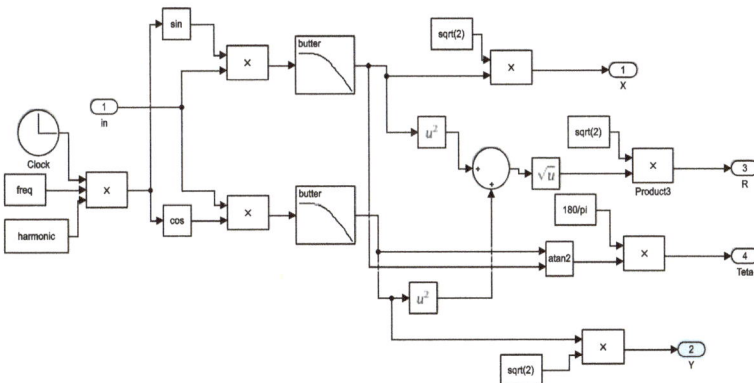

Fig. (2-26). The Simulink model of an analog Lock-In Amplifier.

The Fig. **(2-26)** comes form the Simulink model of [15].

2) A boxcar integrator

Sampling integrator (alternative names are a boxcar average, gated integrator and boxcar integrator) is an electronic instrument that can extract the repetitive signal waveform from the random noise. It uses sampling integration technology, *i.e.* using a very narrow sampling pulse synchronized in the different cycles of signal, and then the noise can be eliminated.

Sampling integrator is an electronic instrument that can extract the repetitive signal waveform from the random noise. Sampling integration technology is adopted, that is, using very narrow sampling pulse synchronized with the signal to sample the same point of input signal buried in the noise, and then calculating the co-correlation between the noise and the signal, in the end, the sampled sample is accumulated synchronously and is outputted through RC low-pass filter [9].

Its principle can be described using formula (2-31) and (2-32).

$$-\sum_{k=0}^{n-1} x(t_0 + kT) = \frac{1}{n}\sum_{k=0}^{n-1} s(t_0 + kT) + \frac{1}{n}\sum_{k=0}^{n-1} n(t_0 + kT) \qquad \textbf{(2-31)}$$

As $\frac{1}{n}\sum_{k=0}^{n-1} n(t_0 + kT) = 0$ so

$$u_{0=}\frac{1}{n}\sum_{k=0}^{n-1} s(t_0 + kT) \cong s(t_0) \qquad \textbf{(2-32)}$$

An example of boxcar integrator device is shown in Fig. (**2-27**), the Matlab simulation is shown in Fig. (**2-28**).

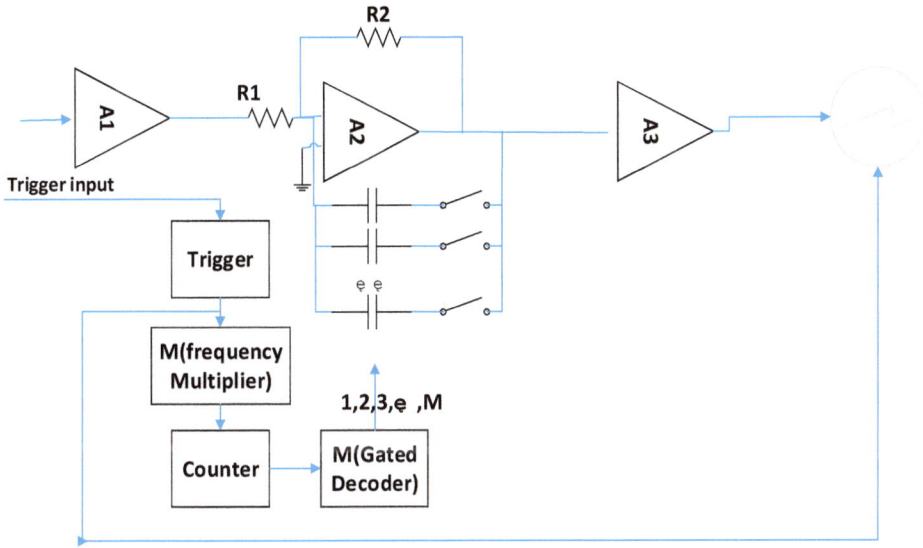

Fig. (2-27). A classic analog boxcar integrator circuit.

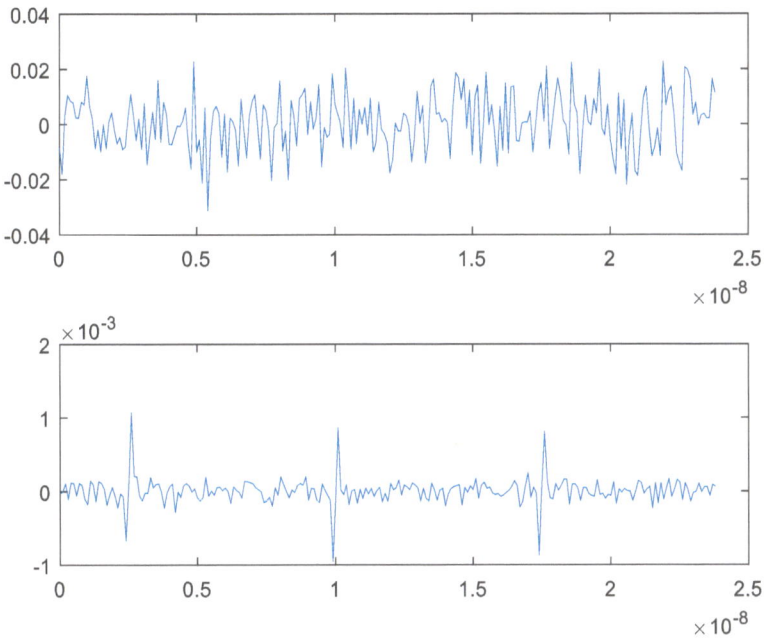

Fig. (2-28). A simulation example of boxcar integrator circuit.

3) Adaptive filter and Noise Canceler

The principle of Adaptive filters is shown in Fig. (**2-29**). In the figure, x(k) represents the input signal value at k time, y(k) represents the output signal value at k time, d(k) represents the reference signal value or the expected response signal value of k, and the error signal e(k) is the difference between d(k) and y(k). The filtering parameters of the adaptive digital filter are controlled by the error signal e (k) and are adjusted automatically according to the value of e(k) to make it suitable for the input x(k+1) at the next time so that the output y(k+1) is close to the desired reference signal d(k+1) [16].

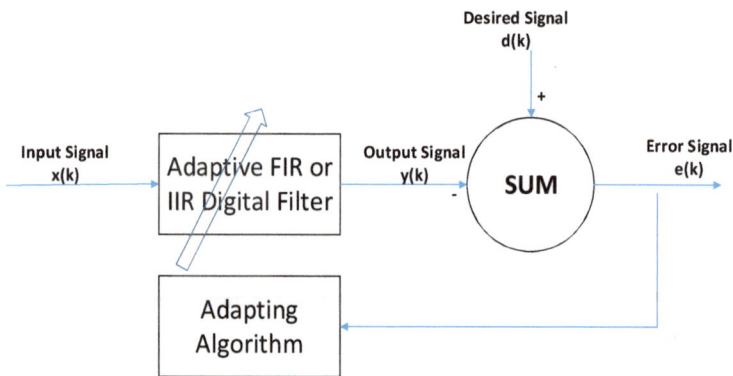

Fig. (2-29). Block Diagram Defining General Adaptive Filter Algorithm Inputs and Outputs.

The research of adaptive filtering algorithm is one of the most active research topics in adaptive signal processing nowadays. Adaptive filtering algorithms are widely used in many fields such as system identification, echo cancellation, adaptive spectral line enhancement, adaptive channel equalization, speech linear prediction, adaptive antenna array and so on. Now, the research of adaptive filtering is to study algorithms with fast convergence speed, low computational complexity and good numerical stability.

A method proposed by Widelow can solve the coefficients of adaptive filters in real time, and the results are close to the approximate solution of Wiener-Hopf equation. This algorithm is called the least mean square algorithm(LMS). This algorithm uses the steepest descent method.

Usually, the ratio of super-mean square error to minimum mean square error (*i.e.* misalignment) is used to evaluate the performance of adaptive filtering.

A. Noise Cancellation

The principle of noise cancellation is shown in Fig. (**2-30**), in noise cancellation, the input signal is n'(k), it is a noise signal related to the wanted to remove noise [16].

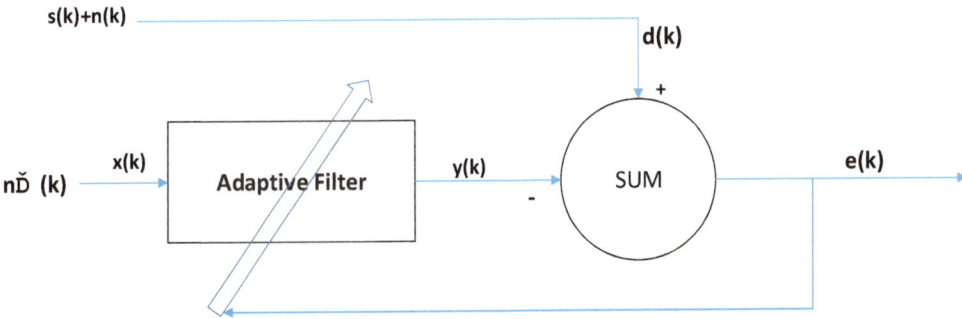

Fig. (2-30). Using an Adaptive Filter to Remove Noise from an Unknown System.

For the x(k) is correlated to y(k) and d(k), in the other way, the input n'(k) can produce the d(k) signal using the filter, this task is fulfilled by the adaptive filter. It adjusts its coefficients to reduce the value of the difference between *y(k)* and *d(k),* then it cancellate the noise, and output the clean signal *e(k)*.

The First method: The Sign-Data LMS Algorithm

The principle of LMS Algorithm is shown in formula (2-33) and formula (2-34).

In these formulae, the μ *is* the step size, the weight coefficient of the filter is w(k). When the error is positive, this steepest descent method produces the new coefficients that it is previous coefficients plus the error multiplied by the step size μ. In formula (2-34), the function of 'sgn' represents this situation.

$$w(k + 1) = w(k) + \mu e(k)sgn[x(k)] \qquad (2\text{-}33)$$

$$sgn[x(k)] = \begin{cases} 1, x(k) > 0 \\ 0, x(k) = 0 \\ -1, x(k) < 0 \end{cases} \qquad (2\text{-}34)$$

With vector *w* containing the weights applied to the filter coefficients and vector *x* containing the input data. The *e(k)* (equal to desired signal - filtered signal) is the error at time *k* and is the quantity the SDLMS algorithm seeks to minimize. The μ (mu) is the step size.

The example of noise cancellation is shown in Fig. (**2-31**).

The sign-sign LMS algorithm is part of the international CCITT standard for 32

Kb/s ADPCM telephony.

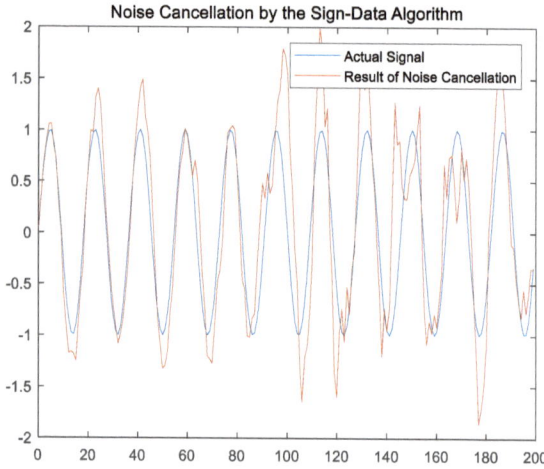

Fig. (2-31). Noise cancellation by the sign-data algorithm.

B. Prediction

In such applications, the role of adaptive filters is to provide a sense of the best prediction of the current value of random signals. Thus, the current value of the signal is used as the desired response of the adaptive filter. The past value of the signal is added to the input of the filter.

Depending on the application of interest, the output or estimation error of the adaptive filter can be used as the output of the system. In the first case, the system acts as a predictor; in the second case, the system acts as a predictive error filter. Its principle is shown in Fig. (**2-32**).

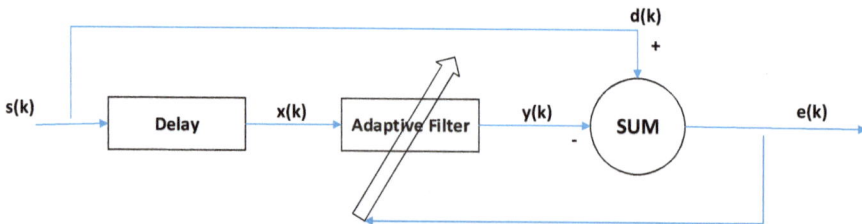

Fig. (2-32). Predicting Future Values of a Periodic Signal.

4) Chaos

Chaos must have these properties [17]:

- sensitive to initial conditions,

- be topologically transitive,

- have dense periodic orbits.

Duffing equation is a chaotic system:

$$x''(t) + kx'(t) - x^3(t) + x^5(t) = f\cos(t) \qquad \textbf{(2-34)}$$

Where k represents the damping ratio; f represents the amplitude of the cycle driving motivation. It can be rewritten using formula 2-35.

$$x'' = -\omega kx + x\omega^2[x - x^3 + f\cos(\omega t)] \qquad \textbf{(2-35)}$$

Where,ω represents the frequency of cycle driving force.

Chaos refers to the unpredictable and stochastic motions of deterministic dynamical systems that are sensitive to initial values. Also known as chaos.

The determinacy of a dynamic system is a mathematical concept, which means that the state of the system at any time is determined by the initial state. Although the motion state at any future time can be deduced from the initial state data and the motion law of the motion, the predicted results will inevitably be erroneous or even unpredictable because the determination of the initial data cannot be completely accurate. The predictability of motion is a physical concept. Even if a movement is deterministic, it can still be unpredictable. Chaos refers to a seemingly irregular complex movement pattern in the real world. The common feature is that the original orderly motion pattern follows the simple physical law, which suddenly deviates from the expected regularity under certain conditions and becomes disordered. Chaos can occur in a wide range of deterministic dynamical systems. Chaos is like stochastic processes in statistical properties and is considered as an intrinsic stochastic property in deterministic systems.

The Matlab code of Duffing function is in below:

```
function y=Duffing(t,x)

global Gamma;

y(1)=x(2);

y(2)=x(1)-0.1*x(2)-(x(1))^3+Gamma*cos(1.25*t);
```

y=[y(1);y(2)];

5) Stochastic resonance and weak signal detection

The existence of noise reduces the signal-to-noise ratio and affects the extraction of useful information. However, in some specific nonlinear systems, the existence of noise can enhance the detection ability of weak signals. This phenomenon is called stochastic resonance.

From signal processing, in a non-linear system, when the noisy signal is input, the system characteristics, such as signal-to-noise ratio, dwell time and so on, are measured by appropriate physical quantities. By adjusting the input noise intensity or system parameters, the system characteristics can reach a maximum value. In this case, we call it signal, noise and non-linear follow-up. The synergistic phenomenon of mechanical system is stochastic resonance.

The function of double-well potential is $V(x) = \left(\frac{1}{4}\right)bx^4 - \left(\frac{1}{2}\right)ax^2$ its

minimum points are located at $\pm x_m$, where $x_m = \left(\frac{a}{b}\right)^{1/2}$. The height of potential barrier is $\Delta V=a2/(4b)$. The top of the potential barrier is located at xb=0.

The weak signal is a periodic driving signal A0xcos(Ωt). When it works, it is running under the potential barrier, in an antisymmetric manner [18]. The Fig. (2-33) is a symmetric double well.

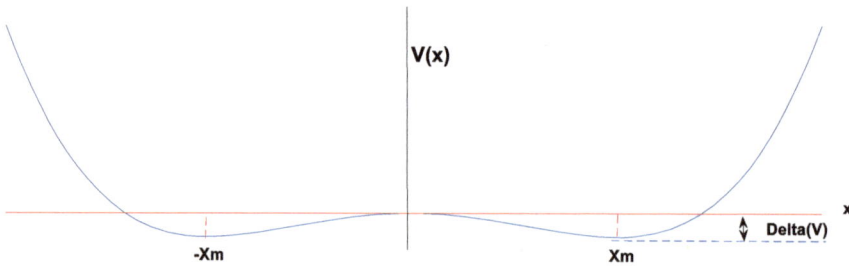

Fig. (2-33). A symmetric double well.

The noise can make the particle roll periodically from one potential well into the other one. It is in some case, the periodic frequency is like weak periodic forcing.

For example, if the average waiting time TKD=1/rk between two noise-induced interwell transitions is comparable with half the period TΩ of the periodic forcing.

$$2T_K(D) = T_\Omega \tag{2-36}$$

Consider the overdamped motion of a Brownian particle in a biostable potential in the presence of noise and periodic forcing.

$$\dot{x} = -V'(x) + A_0 \cos(\Omega t + \varphi) + \xi(t) \tag{2-37}$$

In the absence of periodic forcing, x(t) fluctuates around its local stable states with a statistical variance proportional to the noise intensity D. Noise-induced hopping between the local equilibrium states with the Kramers rate.

$$r_k = \frac{1}{\sqrt{2\pi}} \exp(-\frac{\Delta V}{D}) \tag{2-38}$$

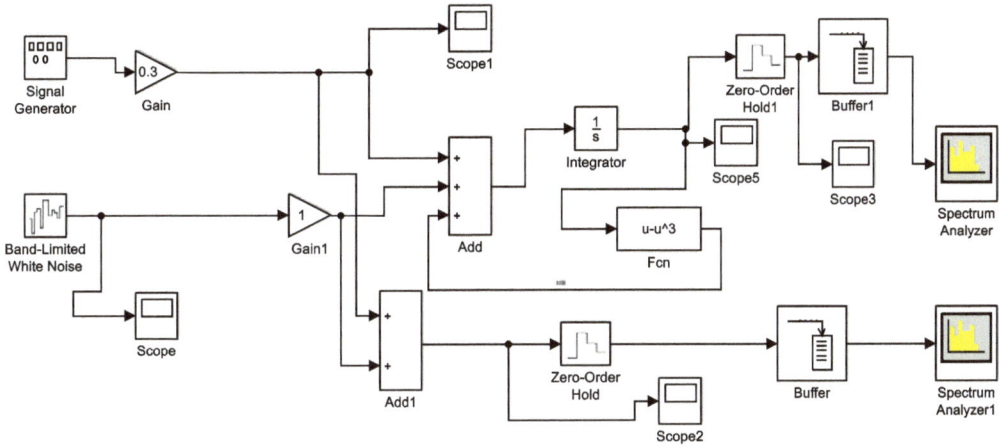

Fig. (2-34). Simulink model for Stochastic resonance [7].

The MATLAB code is shown in following [7]:

```
clear all

clc

fs=5;

f0=0.01;

Ts=1/fs;

h=1/fs;
```

```
t=0:Ts:4095*Ts;

D=0.6;

a=1;

b=1;

s=0.3*sin(2*pi*0.01*t);

p=s+sqrt(2*D)*randn(size(t));

y=fft(s,4096);

pyy=y.*conj(y)/4096;

ff=fs*(0:2048)/4096;

Fig. (1);

subplot(2,1,1);

plot(t,s);

title('input no noise signal');xlabel('time t/s');

ylim([-0.5,0.5]);ylabel(amplitude A');

subplot(2,1,2);plot(ff,pyy(1:2049));

xlabel('Frequency f/Hz');

ylabel('frequency amplitude');xlim([0,0.05]);

title('The Frequency response of no noise signal');

y=fft(p,4096);

pyy=y.*conj(y)/4096;

ff=fs*(0:2048)/4096;

Fig. (2);

subplot(2,1,1);

plot(t,s);
```

```
title(' input no noise signal ');xlabel(' time t/s ');

ylim([-0.5,0.5]);ylabel(' amplitude A ');

subplot(2,1,2);plot(ff,pyy(1:2049));

xlabel(' Frequency f/Hz ');

ylabel(' frequency amplitude ');xlim([0,0.05]);ylim([0,1500]);

title(' The Frequency response of no noise signal ');

x=sr(a,b,h,p);

y=fft(x,4096);

py=y.*conj(y)/4096;

ff=fs*(0:2048)/4096;

Fig. (3);

subplot(2,1,1);

plot(t,x);

title('output signal');

xlabel(' time t/s ');

ylabel(' amplitude A ');

subplot(2,1,2);plot(ff,pyy(1:2049));

xlabel(' Frequency f/Hz ');

ylabcl(' frequency amplitude ');

xlim([0,0.05]);ylim([0,1500]);

title(' The Frequency response of output signal ');

% ***********************************************

function x=sr(a,b,h,x1)

x=zeros(1,length(x1));
```

```
for i=1:length(x1)-1

k1=h*(a*x(i)-b*x(i).^3+x1(i));

k2=h*(a*(x(i)+k1/2)-b*(x(i)+k1/2).^3+x1(i));

k3=h*(a*(x(i)+k2/2)-b*(x(i)+k2/2).^3+x1(i+1));

k4=h*(a*(x(i)+k3)-b*(x(i)+k3).^3+x1(i+1));

x(i+1)=x(i)+(1/6)*(k1+2*k2+2*k3+k4);

end
```

PROBLEMS

2-1 In data acquisition system, the preamplifier should be placed before or after the filter. Why? Write out the derivation process.

2-2 Brief description of the process of designing an active filter circuit

2-3 Brief description of the principle of successive approximation AD converter (combined with ADC0809)

2-4 Brief introduction to the principle of Dual-Integral AD converter combined with ICL7135

REFERENCES

[1]   "Multiplexing" available from https://en.wikipedia.org/wiki/ [Accessed: 2019.2.1].

[2]   "Sample and hold" available from https://en.wikipedia.org/wiki/ [Accessed: 2019.2.1].

[3]   "Analog-to-digital converter" available from https://en.wikipedia.org/wiki/ [Accessed: 2019.2.1].

[4]   "Successive... ADC" available from https://en.wikipedia.org/wiki/Successive_approximation_ADC [Accessed: 2019.2.1].

[5]   "Analog-to-digital converter" available from https://en.wikipedia.org/wiki/ [Accessed: 2019.2.1].

[6]   "Integrating ADC" available from https://en.wikipedia.org/wiki/Dual-slope [Accessed: 2019.2.1].

[7]   L. Fu, and C.J. Deng, "Theory, design and application of intelligent instruments. China, Chengdu: Southwest Jiao tong University Press,2014",

[8]   B.G. Goldshtein, "A Central Limit Theorem of "Non-Commutative" Probability Theory", *Theory Probab. Appl.,* vol. 27, no. 4, pp. 657-665, 2006.

[9]   "Noise" available from https://en.wikipedia.org/wiki/ [Accessed: 2019.2.1].

[10]   Abhirup Lahiri. Simulation of Color Noise (https://www.mathworks.com/matlabcentral/fileexchange/ 18269-simulation-of-color-noise), MATLAB Central File Exchange. Retrieved May 23, 2020.

[11]   D.F Chen, Q Lin, The intelligent instrument, China, Beijing: The China Machine Press.2014.

[12]   Fischer, J. H. "Noise sources and calculation techniques for switched capacitor filters." Solid-State

Circuits, IEEE Journal of 17.4(1982):742-752.

[13]   Walt Kester, Analog-Digital Conversion, Analog Devices, 2004, ISBN 0-916550-27-3, Chapter 3. Also available as The Data Conversion Handbook, Elsevier/Newnes, 2005, ISBN 0-7506-7841-0, Chapter 3.

[14]   Hong-Mei, M. A. "Optoelectronic techniques for calibrating the risetime of a sampling oscilloscope." Journal of Astronautic Metrology & Measurement (2008).

[15]   Predrag Drljaca. Lock in Amplifier (https://www.mathworks.com/matlabcentral/fileexchange/1902-lock-in-amplifier), MATLAB Central File Exchange. Retrieved May 23, 2020.

[16]   Haykin, Simon. Adaptive Filter Theory, 4e. Adaptive Filter Theory. 2002.

[17]   Gleick, James. Chaos: making a new science. 1987.

[18]   Gammaitoni, Luca, and A. R. Bulsara. Stochastic resonances in underdamped bistable systems. Stochastic Processes in Physics, Chemistry, and Biology. 2000.

## CHAPTER 3

The Data Processing Technology of Intelligent Instruments

Abstract: In this chapter, we will discuss the measurement uncertainty in intelligent instruments; the data processing algorithms in industrial intelligent instruments; the inverse problem and its processing method; the intelligent computing includes the deep learning and machine learning arithmetic in the intelligent instrument design.

Keywords: CNN, DAS channel, Deep learning, FFT, Inverse problem, Machine learning, Neural Networks, PID, Software filter, SPWM, Uncertainty.

3.1. THE MEASUREMENT UNCERTAINTY IN INTELLIGENT INSTRUMENTS

Measurement uncertainty is a parameter associated with measurement results, which is used to characterize the dispersion of the values reasonably assigned to the measurement. It can be used in "uncertainty" mode, or it can be a standard deviation (or a given multiple) or the half-width of a given confidence interval [1].

Measuring uncertainty is understood from the word meaning, which means the degree of doubt or uncertainty about the reliability and validity of measurement results. It is a parameter to quantitatively explain the quality of measurement results. In fact, due to imperfect measurement and people's lack of understanding, the measured values are dispersive, that is, the results measured are not the same value, but scattered in a certain area with a certain probability of many values. Although the objective systematic error is an invariant value, because we cannot fully understand or master it, we can only think that it exists in a certain region with a certain probability distribution, and the probability distribution itself is decentralized. Measuring uncertainty is a parameter that indicates the dispersion of the measured value. It does not indicate whether the measured result is close to the true value.

To characterize this dispersion, measurement uncertainty is expressed by a standard deviation. In practice, it is often desirable to know the confidence interval of measurement results. Meanwhile, the uncertainty of measurement can also be expressed by the multiple of standard deviation or the half-width of the interval indicating the confidence level. In order to distinguish the two different representations, they are called standard uncertainty and extended uncertainty respectively.

The standard uncertainties obtained by evaluating the statistical analysis of observation are called Class A standard uncertainties. The standard uncertainties evaluated by statistical analysis of observation columns are called Class B standard uncertainties. Class A uncertainty is derived from a set of probability density functions derived from the observed frequency distribution. Class B uncertainty is based on the degree of trust in an event. They are all based on probability distribution and are characterized by variance or standard deviation. There is no more reliable problem of the two kinds of uncertainty. The reliability of Class A uncertainty depends on the independence of measurement, whether it is in the state of statistical control and the number of measurements.

There is no simple correspondence between the uncertainty of "A" and "B" and the classification of "random error" and "systematic error". "Random" and "system" represent two different properties of error, and "A" and "B" represent two different evaluation methods of uncertainty. The combination of random error and systematic error has no definite principle to follow, which leads to the difference and confusion in the processing of experimental results. The standard uncertainty is used for the combination of Class A uncertainty and Class B uncertainty, which is also one of the advances of uncertainty theory.

1) DAS channel design

Example 3-1: 8-way analog input (alternating signal, f=100Hz), voltage range of 0~10V, the conversion time is less than 50µs, resolution 2mV, channel error is less than 0.1%. The demand is to design the suitable devices and their parameters in these channels [2].

Solve 3-1:

A. The typical DAS channel includes MUX, SH, ADC. First determine ADC: its type, its conversion bit, and then the example device.

The conversion time should follow the Shannon Sampling principle. For the 8 channel 100Hz, the Nyquist frequency is larger than 1600Hz, so the acquired time should be smaller than 0.625ms (or C=3, the time is 0.4ms). The given conversion time is less than 50us, so this conversion time is suitable for the example: 8-way analog input (alternating signal, f=100Hz), even in the worst case the channel and data output time is the same as conversion time. For conversion time is less than 50μs, normally the conversion time of the double integral ADC is larger than 20ms. Simply, commercial ADC mainly includes SAR, sigma-delta, and double integral ADC. the SAR type of ADC is selected for its conversion time is almost between 0.1us and 1ms.

The conversion bit is determined by the formula (3.1), its measurement range, that is the voltage range of 0~10V, its resolution 2mV. It should be 13bits.

$$\frac{10000\text{mV}}{2mV} = 5000, and, 2^{13} = 1024 \times 8 \tag{3-1}$$

And it is verified by channel error (less than 0.1%). Its quantization error is

$$\frac{10}{2^{m+1}} < \delta \rightarrow \frac{10}{2^{m+1}} < 0.01 \Rightarrow 2^{m+1} \geq 10000 \Rightarrow m = 13 \tag{3-2}$$

As an example, we chose the 12bits ADC, AD574, its conversion time is 25us.

A. If the SAR type of ADC were selected, the Sample and Hold circuit should be considered that it is needed or not.

The signal change rate is calculated using equation (3.3).

$$\frac{dU_i}{dt}\Big|_{max} = \omega U_m = 2\pi f U_m \tag{3-3}$$

Guaranteed 1LSB accuracy, the maximum amount of change in the converted signal does not exceed 1 quantization units during the conversion time, etc. Then we get (3.4).

$$2\pi f U_m \times t_c \leq q = \frac{U_m}{2^m} \Rightarrow f_{max} = \frac{1}{2^{m+1}\pi t_c} \tag{3-4}$$

If the conversion time of an ADC is 25us, and its precise has 13 bits, its max signal frequency is 0.75Hz; and the 12 bits ADC, its max signal frequency is 1.5Hz. So, in the example, the channel should have S/H circuit. In this case, the converted signal does not exceed 1 quantization unit during the aperture time tap of the Sample and Hold circuit.

$$f_{max} = \frac{1}{2^{m+1}\pi t_{ap}} \tag{3-5}$$

The aperture time tap of the Sample and Hold circuit LF398 is 35ns, its max signal frequency is 550Hz.

Table 3-1. The true table of AD574.

CE	CS	R/C	12/8	A0	Operation
0	X	X	X	X	None
X	1	X	X	X	None
1	0	0	X	0	Initiate 12-Bit Conversion
1	0	0	X	1	Initiate 8-Bit Conversion
1	0	1	Pin1	X	Enable 12-Bit Parallel Output
1	0	1	Pin15	0	Enable 8 Most Significant Bits
1	0	1	Pin15	1	Enable 4 LSBs +4 Trailing Zeroes

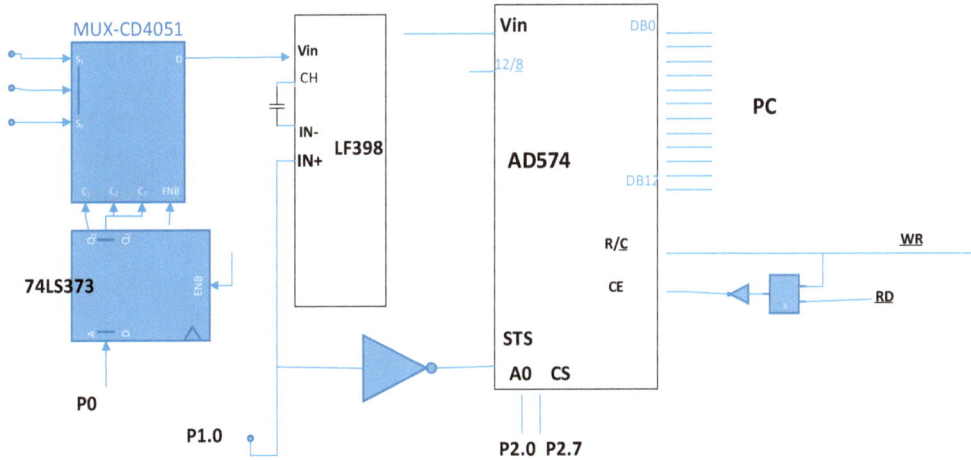

Fig. (3-1). The Schematic design of a channel of DAS.

B. The choice of the other device

Multiplex Analog Switch: Choose 8-way analog switch CD4051, its switching leakage current is about 0.08nA, when the signal source internal resistance is 10kΩ, the error voltage is 0.8μV, it can be negligible. Its switch connection resistance is about 200Ω, and the input resistance of the sampling actuator is generally above 10MΩ, So,when the maximum voltage of supply power is 10V, the voltage drop on the switching resistor is only 0.2mV, then it is negligible also. Fig. (**3-1**) is drawn with reference [2].

The A/D conversion code is

TRANS: MOV R0,#7CH

MOVX @R0, A; start A/D conversion

LOOP: JB P1.0, LOOP; judge the stop of conversion or not

MOV R0,#7DH ; output high 8 bits

MOVX A,@R0 ;

MOV R2,A ;store high 8bits in R2

MOV R0,#7FH ;output low 4bits

MOVX A,@R0 ;

MOV R3,A ;

2) Remote Temperature test channel design.

Example 3-2: Analog Input Channel: The known full test range of temperatures is 100 degrees, temperatures in the 16 or 8 working places should be measured, the total error of the analog input channel is required to be ±1.0 degree, the ambient temperature is 25±15°C, and the power supply fluctuates about ±1% [2].

Solve 3-2:

A. The uncertainty measurement models.

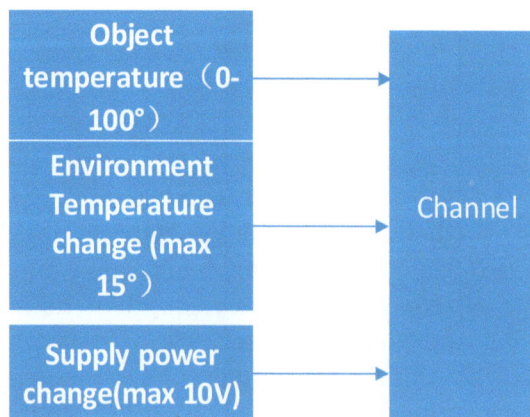

Fig. (3-2). The measurement model.

B. The uncertainty(error) disperse

For the temperature normally change very small, the structure of multi-channel sharing the same SH and A/D converter can be selected.

Since the sensor and signal amplification circuit are the main parts of the channel, the main part of the error is assigned to these parts, nearly (90%) ±0.9 degrees. The other parts are assigned 10%, including A/D conversion, multiplex switching, sampling to maintain ± 0.1 degrees.

C. The Sensor and analog amplify circuit design, Fig. (3-3) is redrawn with reference [2].

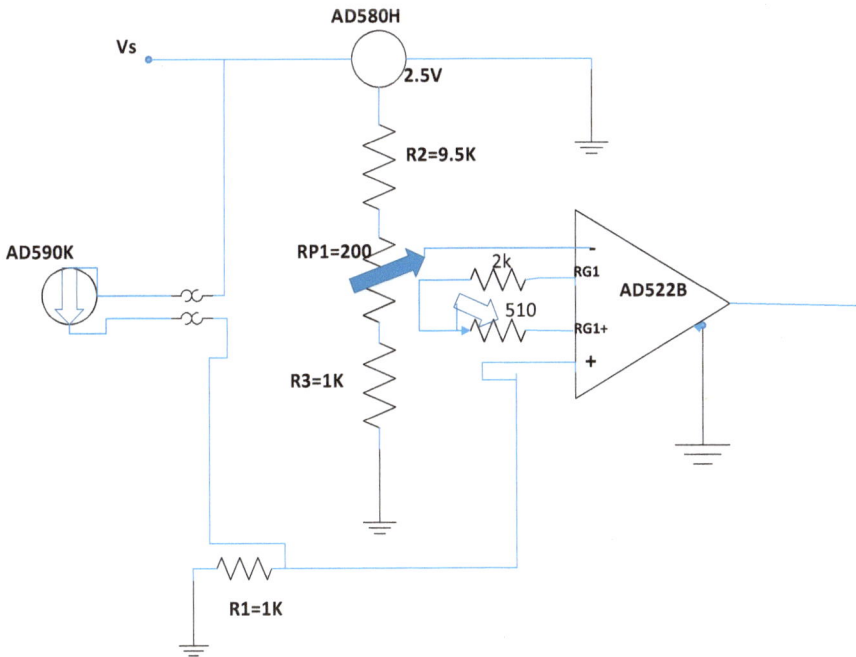

Fig. (3-3). The temperature test circuit of a single channel.

The error estimations of the sensor and condition circuit are presented below:

Sensor Error has the following parts: 1) AD590 linear error; 2) power supply suppression error; 3) temperature error of current voltage transform resistance.

Signal Amplifier mainly has the following error: 1) temperature error of AD580;

2) resistance voltage temperature error ;3) the common-mode error, offset voltage temperature drift error, gain temperature coefficient error, the linear error of AD522.

The estimation of Class B measurement uncertainty of sensor:

A) Temperature error of AD590: Linear error of AD590 is 0.20°

B) AD590K Power suppression error, when between +5V and +15V, its suppression ratio is 0.2°/V, if the power supply is 10V, it is 0.02° error with 0.1V change.

C) Resistance voltage temperature error Resistance 1K, resistance error 0.1%, the temperature coefficient is 25×10^{-6}°C. The AD590 sensitivity is that $1\mu A/$°C, the 0° output 273.2uA. When the environment changes 15°,

$$273.2 \times 10^{-6} \times 10 \times 10^{-6} \times 15 \times 10^3 = 0.04mV(0.04°C)$$

The estimation of Class B measurement uncertainty of the signal amplifier circuit:

$$273.2 \times 10^{-6} \times 25 \times 10^{-6} \times 15 \times 10^3 = 0.1mV(0.1°C)$$

D) AD580, resulting in 273.2uA*1k bias, voltage temperature coefficient 25×10-6/°C

$$273.2 \times 10^{-6} \times 10 \times 10^{-6} \times 15 \times 10^3 = 0.04mV(0.04°C)$$

E) Resistance voltage, resistance R2, R3, the temperature coefficient is Bias voltage variation 25×10^{-6}°C.

$$273.2 \times 10^{-6} \times 10 \times 10^{-5} \times 10^3 = 0.0027mV$$

F) A common-mode error of AD522, with a gain of 100, CMRR 100dB and a common-mode voltage of 273.2mV

G) AD522 offset voltage temperature drift error, offset voltage temperature coefficient $2.5 \times \frac{10^{-6}\mu V}{°C}$

$$(2.25) \times 10^{-6} \times 15 = 0.03mV(0.03°C)$$

H) Gain temperature coefficient of AD522 AD522 $25 \times \frac{10^{-6}}{°C}$.   Resistance gain temperature coefficient $25 \times 10^{-6} °C$. When the $15°$ changes,

$(25 + 10) \times 10^{-5} \times 15 \times 100 = 0.05mV(0.05°C)$

F) AD522 Linear error 0.002%, (gain 100) in output 10V

  $0.00002 \times 10 = 0.2mV(0.2°C)$

Calculation:

Absolution SUM: $0.68°$

Square root sum: $0.31°$

D. The design of A/D channel

The error estimation of other parts: A/D conversion device selects AD5420BD (linear and quantitative error), MUX selects AD7501, and normal S/H.

The errors include normal linear and quantitative error of AD5420BD, leakage voltage, the linear error of S/H, and their temperature effects: Influence of S/H offset drift, gain drift and power supply voltage change A/D offset drift, gain drift, and so on.

A) The estimation at room temperature

Multi-channel switch leakage current is $10 \times 10-9 \times 8 = 0.08 \mu V$

MUX input signal ratio is $300 \div 1011 = 3nV$

S/H Linear Error is 0.01%

A/D linear error is 0.012%; quantization 13 bits, quantization error is 0.012%; aliasing error is 0.01%

Error Absolute value sum is $0.044°C$

Error Square Root sum is $0.022°C$

B) Temperature range

S/H offset temperature Drift error is 0.015%, it's gain offset error is 0.015%

A/D gain offset error is 0.037%; offset temperature drift error is 0.01%; power supply voltage offset is 0.009%;

Error Absolute value sum is 0.086°C

Error Square Root sum is 0.045°C

3.2. THE DATA PROCESSING ALGORITHMS OF INTELLIGENT INSTRUMENTS

3.2.1. The Data Filter Arithmetic

1) The limit value data filter

For random interference, limit filtering is an effective method. Basic method: Compare two sampling values $Y(n)$ and $Y(n-1)$ for adjacent n and n-1 moments, and determine the maximum deviation allowed for two samples based on experience. If the difference of two sampling values exceeds the maximum deviation range, it is considered that random interference occurs and the later sampling value $Y(n)$ is considered to be illegal and should be deleted, after deleting $Y(n)$, $Y(n-1)$ can be used instead of $Y(n)$, and this sampling value is considered valid if the maximum deviation range allowed is not exceeded.

Example 3-3:

```
#define A 30

unsigned char value;

unsignech char filter()

{ unsigned char new_value;

new_value = get_ad();

if ((new_value - value > A) || (value - new_value > A)) return value;

return new_value;

}
```

2) The median value data filter

Median filtering method can effectively overcome the fluctuation caused by accidental factors or the error caused by sampling instability and other pulse interference. This method can be used to get a good filtering effect on the measured parameters such as temperature and level, but it is generally not appropriate to use the median filtering method for fast changing parameters such as flow pressure. Basic method: Continuous sampling n times (usually n odd numbers) for a measured parameter, and then the sampling values are arranged by value, in the end the median value is outputted as the sample value.

Example 3-4:

#define K 17

unsigned char filter()

{ unsigned char value_buf[K], cnt,i,j,temp;

for (cnt=0;cnt<K;cnt++)

{ value_buf[cnt] = get_ad(); delay(); }

for (j=0;j<K-1;j++)

{ for (i=0;i<K-j;i++)

{ if (value_buf[i]>value_buf[i+1])

{tmp = value_buf[i]; value_buf[i] = value_buf[i+1]; value_buf[i+1] = tmp; }

}

}

return value_buf[(K-1)/2];

}

3) The average value data filter

The arithmetic average filtering method is suitable for filtering the general signal with random interference.

The characteristic of this signal is that the signal itself fluctuates up and down near a certain range of values, such as measuring flow rate and liquid level;

Basic method: calculating the average input of N sampling data.

Example 3-5:

#define K 20

unsigned char filter()

{unsigned int sum = 0,count;

for (count=0;count<K;count++)

{ sum+=get_ad(); delay();}

return (unsigned char)(sum/N);

}

4) The moving average data filter

Basic method: The array is used as the measurement data memory, the length of the array is N, each new measurement is placed at the end of the array, and withdraw the first data from the array, so that there is always N "up-to-date" data in the array. When the average value is calculated, a new arithmetic average can be obtained by averaging the N data in the new array. In this way, a new arithmetic average can be obtained for each measurement.

Example 3-6:

#define K 22

unsigned char value_buf[K],i=0;

unsigned char filter()

{ unsigned char count; int sum=0;unsigned char tmp,j;

tmp= get_ad();

//

for(j=0;j<K-1;j++)

value_buf[i]=value_buf[i++];

value_buf[K]=tmp;

if (i == K) i = 0;

for (count=0;count<K;count++)

sum = value_buf[count];

return (char)(sum/K);

}

5) The first-order latency data filter

It is good to decrease the effect of periodic interference and is suitable for the measurement with high sample frequency.

Disadvantages: its Phase lag, and its low sensitivity. The latency depends on the 'a' value size. The interference signal with a filter frequency higher than 1/2 of the sampling frequency cannot be eliminated. The procedure is as follows:

Example 3-7:

#define b 50

unsigned char value;

unsigned char filter()

{unsigned char new_value;

new_value = get_ad();

return (100-b)*value + b*new_value;

}

3.2.2. Fast Fourier Transform Arithmetic

Fast Fourier Transform(FFT) is a fast and effective algorithm to reduce the number of discrete Fourier transform (DFT) calculations. It is based on the singularity, virtual and real characteristics of discrete Fourier transform, and the algorithm of discrete Fourier transform is improved.

The basic idea of FFT transforms algorithm: Using the periodicity and symmetry of W_N, an N-item sequence (set $N=2k$, k as a positive integer) is divided into two N/2 sub-sequences, each N/2 point DFT transformation Needs $(N/2)^2$ Sub-operation, and then N operations to two N/2, The 2-point DFT transform is combined into an N-point DFT transform. After this transformation, the total number of operations becomes $N+2(N/2)^2=N+N^2/2$.

Example 3-8:

```c
#include <stdio.h>

#include <stdlib.h>

#include <math.h>

typedef struct{

double r;

double i;

}my_complex;

#define NOT2POW(a) (((a)-1)&(a)||(a)<=0)

#define MYPI 3.14159265358979323846//pi

my_complex* fft(const my_complex* x, unsigned int len){

unsigned int ex=0,t=len;

unsigned int i,j,k;

my_complex *y;

double tr,ti,rr,ri,yr,yi;

if(NOT2POW(len)) return NULL;

for(;!(t&1);t>>=1) ex++;

y=(my_complex*)malloc(len*sizeof(my_complex));

if(!y) return NULL;

for(i=0;i<len;i++){k=i; j=0; =ex;

while((t--)>0){ j<<=1; j|=k&1; k>>=1; }

if(j>=i){ y[i]=x[j]; y[j]=x[i]; }

}

for(i=0;i<ex;i++){t=1<<i;
```

```c
for(j=0;j<len;j+=t<<1){ for(k=0;k<t;k++)

{ti=-MYPI*k/t; rr=cos(ti); ri=sin(ti);

tr=y[j+k+t].r; ti=y[j+k+t].i;

yr=rr*tr-ri*ti; yi=rr*ti+ri*tr;

tr=y[j+k].r; ti=y[j+k].i;

y[j+k].r=tr+yr; y[j+k].i=ti+yi;

y[j+k+t].r=tr-yr; y[j+k+t].i=ti-yi;}}}

return y;}

int main()

{int i,DATA_LEN;

my_complex *x,*y;

printf("BASE-2 FFT TEST\n input length:");

scanf("%d",&DATA_LEN);

x=(my_complex*)malloc(DATA_LEN*sizeof(my_complex));
for(i=0;i<DATA_LEN;i++){x[i].r=i; x[i].i=i-1; }

printf("Preporcessing...\n real\t \t image\n");

for(i=0;i<DATA_LEN;i++)- printf("%lf\t%lf\n",x[i].r,x[i].i);

y=fft(x,DATA_LEN);-if(!y){ printf("length should be based-2\n"); return 0; }

for(i=0;i<DATA_LEN;i++)

printf("%lf\t%lf\n",y[i].r,y[i].i);

free(y); free(x);

return 0;

}
```

3.2.3. Sinusoid Pulse Width Modulation Arithmetic

The full name of PWM is Pulse Width Modulation, which changes the equivalent output voltage by changing the duty-free ratio of the output square wave and is widely used in motor speed regulation and valve control.

SPWM is based on PWM to change the modulation pulse mode, pulse width time duty-free ratio according to Sinusoid wave rate arrangement, so that the output waveform through the appropriate filtering can achieve sine wave output, it is widely used in DC AC inverter and so on [3];

The basis of SPWM theory: When the signal is rebuilt with the narrow equal pulses and different shapes impulse, the almost same waveform signal can be obtained.

1). Equal area method

The scheme replaces the sine wave with the same number of rectangular pulse sequences with unequal width, then calculates the width and interval of each pulse, and saves the data in the microcomputer, and generates the PWM signal control switch device through the method of the looking-up table, to achieve the desired goal. Its advantages are present in the following: The pass time of each switching device can be calculated accurately, and the waveform obtained is very close to the sine wave. Its disadvantages: The calculation is cumbersome, the data consumes a lot of memory, cannot be controlled in real time.

Example 3-9:

void CalcSpwm (float a,unsigned short int w_Hz,unsigned int z_Hz)

{

volatile unsigned short int i,n,*p;

float m,n1,n2;

m = z_Hz/w_Hz ;

g_SPWMTable.SpwmSize =(unsigned short int16)m;

p=g_SPWMTable.p_HeadTable;

n=m; m/=2;

n1=(float)g_T1_Clk/(8.0*m*w_Hz);

n2=(float)g_T2_Clk/(8.0*PI*w_Hz)*a;

for(i=0;i<n;i++)

{*p=n1-n2*(cos(i*PI/m)-cos((i+1)*PI/m)); p++;}

}

2) Hardware modulation method

Program principle: The desired waveform as a modulation signal, the received modulation signal as a carrier, through the modulation of the carrier to obtain the desired PWM waveform. Using isosceles triangular wave as carrier, when the modulated signal wave is sine wave, the SPWM waveform is obtained.

Advantages: The realization method is simple, can solve the complicated shortcoming of equal area method calculation.

Disadvantages: Analog circuit structure is complex; it is difficult to achieve precise control.

3) Natural sampling method

Scheme principle: The sine wave is used as the modulation wave, the isosceles triangle wave is compared for the carrier, and the switching device is controlled at the natural intersection time of two waveforms.

Advantages: The obtained SPWM waveform is closest to the sine wave.

Disadvantages: Pulse width expression is a transcendental equation, the calculation is cumbersome, difficult to control in real time.

4) Rule sampling method

The principle of the Rule Sampling method is to use a triangular wave to sample the sine wave to obtain the ladder wave, and then to control the switching device through the intersection of step wave and triangular wave, to realize the SPWM method. The Rule Sampling method is divided into symmetric rule sampling and asymmetric rule sampling according to the different position of a triangular wave at its vertex or bottom point.

Advantages: Simple calculation, easy on-line real-time operation, in which asymmetric rules sample more than fain number and closer to sinusoidal.

Disadvantages: DC voltage utilization is low, the linear control range is small.

Example 3-10: Symmetrical sampling

```
void CalcSpwm (float a,float w_Hz,float z_Hz)
{unsigned int tmp_PR;
volatile unsigned int i,n,*p;
float m;
m = z_Hz/w_Hz ;
g_SPWMTable.SpwmSize =(unsigned int)m;
tmp_PR = g_T1_Clk /(2*z_Hz);
p=g_SPWMTable.p_HeadTable;
for(i=0;i<(unsigned int)m;i++)
{n=tmp_PR*(0.5-0.5*a*sin((i+0.75)*2*PI/m));
*p=n; p++;}
}
```

Example 3-11: Asymmetrical sampling

```
void CalcSpwm1(float a,unsigned short int w_Hz, unsigned int 32 z_Hz)
{unsigned short int tmp_PR;
volatile unsigned short int i,n,*p;
float m;
m = z_Hz/w_Hz ;
g_SPWMTable.SpwmSize =(unsigned short int)m;
tmp_PR = g_T1_Clk /(2*z_Hz);
p=g_SPWMTable.p_HeadTable;
for(i=0;i<(Uint16)m;i++)
{n=tmp_PR*(0.5-0.25*a*(sin((i+0.25)*2*PI/m)+sin((i+0.75)*2*PI/m)));
```

```
*p=n; p++;}

}
```

3.2.4. PID Arithmetic

In the Process Control, the PID controller (also known as PID regulator), which is controlled by the proportion of deviation (P), Integral (I) and differential (D), is one of the most widely used automatic controllers. For the typical object of Process control, the control object of "first order lag + pure lag" and "second order lag + pure lag", PID controller is an optimal control. PID regulation Law is an effective method for dynamic quality correction of continuous system, its parameter tuning method is simple, and the structure change is flexible (PI, PD...) [4]. Fig. (**3-4**) is redrawn with reference [4].

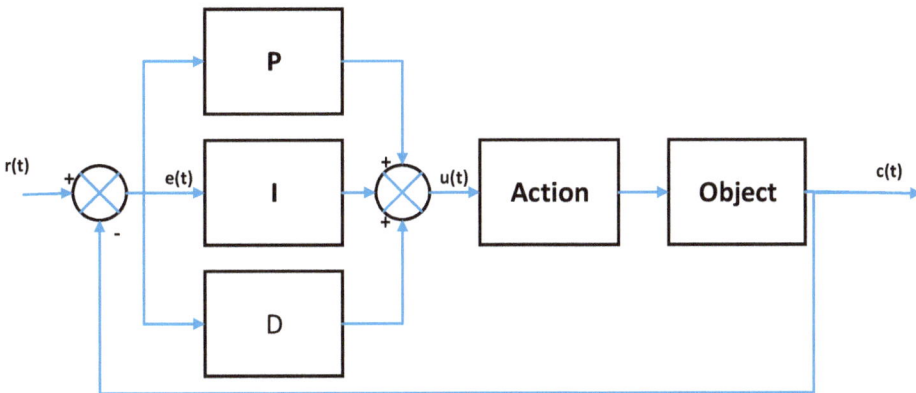

Fig. (3-4). The block diagram of PID controller.

The function of the PID regulator is presented in the following:

Proportional: The deviation signal $e(t)$ of the immediate proportional reaction control system, once the deviation is generated, the regulator immediately produces a control effect to reduce the deviation;

Integral: Mainly used to eliminate the static error. The larger the integral time constant T_I, the weaker effect the integral function has, and the stronger effect the reverse has.

Differential: Can reflect the changing trend of deviation signal (change rate) and can introduce an early valid correction in the system before the value of the deviation signal becomes too large, thus speeding up the action speed of the system and reducing the adjustment time.

The PID regulator is a linear regulator that controls the control object by combining the ratio of the given value R (t) with the actual output value C (t) (P), the Integral (I), and the differential (D) by a linear combination to form the control.

$$u(t) = K_p[e(t) + \frac{1}{T_I} \int_0^t e(t)dt + T_D \frac{de(t)}{dt}] \tag{3-6}$$

Table 3-2. The discrete form of PID controller.

	The analog form	The discrete form
error	$e(t) = r(t) - c(t)$	$e(n) = r(n) - c(n)$
Differential of error	$\dfrac{de(t)}{dt}$	$\dfrac{e(n) - e(n-1)}{T}$
Integral of error	$\displaystyle\int_0^T e(t)dt$	$\displaystyle\sum_0^n e(i)T$

Example 3-12: PID

```c
#include <reg51.h>

#include <string.h>

typedef struct PID {

double SetPoint_Desiredvalue;

double Proportion_Const;

double Integral_Const;

double Derivative_Const;

double Last_Error; //

double Prev_Error; //

double Sum_Error;

} PID;
```

```c
double PID_Calc(PID *pp, double NextPoint)

{

double dError, Error;

Error = pp-> SetPoint_Desiredvalue - NextPoint;

pp->Sum_Error += Error;

dError = Error - pp->Last_Error;

pp->Prev_Error = pp->Last_Error;

pp->Last_Error = Error;

return (pp-> Proportion_Const * Error

+ pp-> Integral_Const * pp->Sum_Error + pp-> Derivative_Const * dError); }

void PIDInit (PID *pp) { memset (pp,0,sizeof(PID));}

double sensor (void)

{…; }

void actuator(double r)

{}

void main(void) {

PID sPID;

double Out;

double In; PIDInit (&sPID);

sPID. Proportion_Const = 0.15;

sPID.Derivative_Const = 0.01;

sPID.SetPoint_Desiredvalue = 400.0;

for (;;) {

In = sensor ();
```

Out = PIDCalc (&sPID,In);

actuator (Out);

}

}

3.3. THE INVERSE PROBLEM AND ITS PROCESSING METHOD

The inverse problem method is relative to the positive problem-solving method. For example, the forward problem is to solve the pressure or velocity distribution for a given geometric shape, while the inverse problem method is to obtain the corresponding geometric shape according to a given objective function, such as pressure or velocity distribution.

Over the past 30 years, the inverse problem of mathematical physics has become one of the fastest growing fields in applied mathematics, which is largely driven by the urgent needs arising from the application of many engineering and technical fields. In practice. Many inverse problems can be reduced to the first kind of operator equation, and some methods of inverse problems, such as pulse spectrum technique (PST) or generalized pulse spectrum technique (GPST) and optimal perturbation method, often solve the first kind of operator equation as a sub-process.

Soft measurement and intelligent instrument often use the design method of the inverse problem. The soft measurement: According to the mathematical relationship between measurable and easy-to-measure process variables and variables to be measured directly, according to some optimization criterion, various calculation methods are used to realize the measurement and estimation of measurement variables by software.

3.3.1. Example of Scale in Intelligent Instrument Design

(linear) Single input single output inverse problem of instrument parameter calculation

$$Y = f(x, a, b) \qquad (3\text{-}6)$$

Forward problem: $x, a, b \Rightarrow Y$

Inverse problem: $Y, a, b \Rightarrow x$ or $Y, x \Rightarrow a, b$, and so on.

It is obvious that the linear forward problem leads to an easy solution to its inverse problem.

Example: 3-13

The relationship between the resistance value and temperature of a temperature measuring thermistor is

$$R_T = \alpha \cdot R_{25°C} e^{\frac{\beta}{T}} = f(T) \tag{3-7}$$

R_T is the resistance value of the thermistor at the temperature of T;

Solution 3-13:

$$\ln R_T = \ln\left(\alpha \cdot R_{25°C} e^{\frac{\beta}{T}}\right) = \ln(\alpha \cdot R_{25°C}) + \frac{\beta}{T} \tag{3-8}$$

$$T = \beta / \left(\ln\left(\frac{R_T}{\alpha \cdot R_{25°C}}\right)\right) \tag{3-9}$$

Example 3-14:

An intelligent temperature measuring instrument uses an 8-bit ADC with a measuring range of 10~100°C, with a digital quantity of 28H after the instrument has been sampled, filtered and nonlinear corrected (that is, the relationship between temperature and digital volume is linear).

Solution 3-14:

At this point, the $A_0=10°C, A_m=100°C, N_m=FFh=255, N_x=28h=40$

$$A_x = \left(\frac{N_x}{N_m}\right)(A_m - A_0) + A_0 = \left(\frac{40}{255}\right)(100 - 10) + 10 = 24.1°C \tag{3-10}$$

So, the solution of the inverse problem should meet (1) existence; (2) unique; (3) stable.

When there is a pathological inverse problem, a regular theory or method is used to solve the inverse problem.

Regularization, which means that in linear algebra theory, the ill-posed problem is usually defined by a set of linear algebraic equations, and this set of equations usually comes from an unsuitable inverse problem with a large number of conditions [7].

A large number of conditions means that rounding errors or other errors can seriously affect the outcome of the problem. Besides, an explanatory definition is given for linear equations Ax=b, when solution x does not exist or the solution is not unique, it is called pathological problem (ill-posed problem).

But in many cases, we need to solve the morbid problem, so how to do it? In cases where the solution does not exist, the solution is to add some conditions to find an approximate solution, and for cases where the solution is not unique, the solution is to add some limitations to narrow the scope of the solution. This method of solving pathological problems by adding conditions or limiting requirements is a regular method:

Through some adjustments or other means, the pathological problem can only have a solution. In this process of adjustment, the technology used is the regular technology, the method is called as the regular method. The standard method for solving linear equations is the least-squares method, that is, the solution of min||Ax-b|2.

But for the pathological linear equation, Gikhonov proposed the use of methods called Gikhonov matrix

$$\min||A_x - b||^2 + ||\Gamma x||^2 \tag{3-11}$$

Other regularization method includes

- A regular method based on variational principle

- A regular method based on spectral analysis

- Iterative method of the iteration regular method

- The Discrete regular method

3.3.2. Methodologies in Intelligent Instrument Design

(1) As far as possible to collect a wealth of measured data, if conditions permit, can be used multi-sensor and its data fusion technology;

(2) Minimize the system of data acquisition, coarse errors, and deal with random errors. (Should have evaluation function);

(3) Minimize the uncertainty of the measurement system and the collected data (should have the evaluation function);

(4) Accurate mathematical and measurement models;

(5) Choosing a regular method;

(6) Choosing intelligent algorithm;

(7) Write code in accordance with software engineering;

(8) The evaluation function should have the inverse problem solution.

3.4. THE INTELLIGENT COMPUTING

Also known as computational intelligence, including genetic algorithm, simulated annealing algorithm, tabu search algorithm, evolutionary algorithm, heuristic algorithm, ant colony algorithm, artificial fish swarm algorithm, particle swarm optimization, hybrid intelligent algorithm, immune algorithm, artificial intelligence, neural network, machine learning, biological computing, DNA computing, quantum computing, *etc.*

Intelligent computing is developed from general computing.

Intelligent computing needs the following key features:

1) Continuous evolution: the ability of self-intelligent management and upgrading;

2) Environmentally friendly: location-independent deployment, seamless connection, and efficient collaboration;

3) Open Ecology: All the upstream and downstream industries can participate extensively to create and share AI dividends;

It is to use advanced IT and CT technologies (chips, architectures, AI, *etc.*) to upgrade IT infrastructure intelligently (intelligent management, online upgrade, and evolution), distribute optimal computing resources intelligently with different business loads, improve the utilization efficiency of IT infrastructure, and optimize the current business computing TCO; secondly, it is the basis of the future AI. The new business form provides abundant and economical computing, and it can also be developed, deployed, used and coordinated anytime and anywhere, lowering the threshold of AI usage, making AI a universal and inclusive computing resource.

3.4.1. The Deep Learning Arithmetic

Deep Learning (DL) is a new research direction in the field of machine learning. It has been introduced into machine learning to make it closer to the original goal of AI (Artificial Intelligence).

Deep learning is the inherent law and representation level of learning sample data. The information obtained during the learning process is very helpful to the interpretation of data such as text, image and sound. Its ultimate goal is to enable machines to have the same analytical learning ability as human beings, and to recognize data such as text, images and sounds. Deep learning is a complex machine learning algorithm, which has achieved much better results in speech and image recognition than previous related technologies.

Deep learning has achieved a lot in search technology, data mining, machine learning, machine translation, natural language processing, multimedia learning, voice, recommendation and personalization technology, and other related fields. The depth learning enables machines to imitate human activities such as audiovisual and thinking, solve many complex pattern recognition problems and makes great progress in artificial intelligence related technologies.

The Convolutional neural networks (ConvNets) are widely used tools for deep learning. They are specifically suitable for images as inputs, although they are also used for other applications such as text, signals, and other continuous responses.

Deep learning is a general term for a kind of pattern analysis method. As far as the specific research content is concerned, it mainly involves three kinds of methods:

(1) Neural network system based on convolution operation, namely convolution neural network (CNN).

(2) Self-coding neural networks based on multi-layer neurons, including self-coding (Auto-encoder) and sparse coding (Sparse Coding), which have attracted wide attention in recent years.

(3) Pre-training the multi-layer self-coding neural network and further optimizing the depth confidence network (DBN) of neural network weights with discriminant information.

Through multi-level processing, the initial "low-level" feature representation is gradually transformed into "high-level" feature representation, and then the complex classification and other learning tasks can be completed with "simple model". Thus, in-depth learning can be understood as "feature learning" or "representation learning".

In the past, when machine learning was used for real tasks, the features of describing samples were usually designed by human experts, which became "feature engineering". As we all know, the quality of features has a crucial impact on generalization performance, and it is not easy for human experts to design good features; feature learning (representation learning) generates good features through machine learning technology itself, which makes machine learning a step forward to "automatic data analysis"

Convolutional Neural Network Model: Before the emergence of unsupervised pre-training, training depth neural network is usually very difficult, and one of the special cases is convolutional neural network. Convolutional nerve network is inspired by the structure of visual system. The first computational model of convolutional neural network was proposed in Fukushima (D) neurocognitive machine. Based on the local connection between neurons and hierarchical tissue image conversion, neurons with the same parameters were applied to different positions of the former layer of neural network to obtain a translation invariant neural network structure. Later, based on this idea, Le Cun *et al.* designed and trained convolutional neural networks with error gradient and obtained superior performance in some pattern recognition tasks. So far, the pattern recognition system based on convolutional neural network is one of the best implementation systems, especially in handwritten character recognition tasks, showing remarkable performance.

Deep Trust Network Model: DBN can be interpreted as Bayesian probabilistic generation model, which consists of multi-layered random hidden variables. The upper two layers have undirected symmetrical connections. The lower layer gets top-down directional connections from the upper layer. The state of the lowest unit is the visible input data vector. DBN consists of a stack of 2F structural units, which are usually RBM (Restricted Boltzmann Machine). The number of visual neurons in each RBM unit in the stack is equal to the number of hidden neurons in the previous RBM unit. According to the deep learning mechanism, the first layer RBM unit is trained with input samples, and the second layer RBM model is trained with its output. The RBM model is stacked by adding layers to improve the performance of the model. In the unsupervised pre-training process, after DBN coding is input to the top-level RBM, the top-level state is decoded to the bottom-level unit, and the input is reconstructed. RBM, as the structural unit of DBN, shares parameters with each layer of DBN.

Stack self-coding network model: The structure of the stack self-coding network is similar to that of DBN. It consists of several stacks of structural units. The difference is that the structural units are auto-encoder rather than RBM. The self-coding model is a two-layer neural network. The first layer is called the coding layer and the second layer is called the decoding layer.

In 2006, Hinton proposed an effective method to build a multi-layer neural network on unsupervised data, which is divided into two steps: first, to build a single layer neuron layer by layer, so that each time a single layer network is trained; when all layers are trained, the wake-sleep algorithm is used to optimize.

In this way, the top layer is still a single layer of neural network, while the other layers are transformed into graph models. The upward weight is used for "cognition" and the downward weight is used for "generation". Then the wake-sleep algorithm is used to adjust all the weights. To achieve agreement between cognition and generation is to ensure that the top-level representation generated can restore the bottom nodes as accurately as possible. For example, if a node at the top represents a face, then all face images should activate the node, and the resulting downward generated image should be able to represent an approximate face image. Wake-sleep algorithm is divided into wake and sleep.

Wake stage: cognitive process, which generates abstract representations of each layer through external characteristics and upward weights, and uses gradient descent to modify downward weights between layers.

Sleep stage: Generation process, through the top representation and downward weights, generates the state of the bottom, while modifying the weights between layers up.

Unsupervised Learning from Down to Up: It's from the bottom to the top. This step can be regarded as an unsupervised training process, which is the most distinct part of the traditional neural network and can be regarded as a feature learning process. Specifically, the first layer is trained with uncalibrated data, and the parameters of the first layer are trained first. This layer can be regarded as a hidden layer of a three-layer neural network that minimizes the difference between output and input. Due to the limitation of model capacity and sparseness, the obtained model can learn the structure of the data itself. After learning the n-l layer, the output of the n-l layer is used as the input of the n layer, and the N layer is trained to get the parameters of each layer.

Top-down supervised learning: It is through labeled data to train, error transmission from top to bottom, fine-tuning the network. Based on the parameters obtained in the first step, the parameters of a multi-layer model are further optimized. This step is a supervised training process. The first step is similar to the stochastic initialization of neural networks. Because the first step is not stochastic initialization but is obtained by learning the structure of input data, the initial value is closer to the global optimum and thus can achieve better results. So, the good effect of deep learning is largely attributed to the process of feature learning in the first step.

Example 3-15:

camera = webcam; % Connect to the camera

net = alexnet; % Load the neural network

while true

im = snapshot(camera); % Take a picture

image(im); % Show the picture

im = im resize(im,[227 227]); % Resize the picture for alexnet

label = classify(net,im); % Classify the picture

title(char(label)); % Show the class label

drawnow

end

3.4.2. The Neural Networks

Neural networks are machine models that simulate the structure and working patterns of the human brain. It is a kind of nonlinear and adaptive information processing with a large number of unit interconnections with memory and information processing functions and has a system similar to human intelligence [9].

Artificial nerve element is the basic information processing unit of artificial neural network operation, and it is also the foundation of neural network. Artificial neural models consist of 4 basic elements:

(1) The strength of a set of connections. Connections is represented by the weights on each connection, the weights can take positive values or negative values, the positive weights represent activation, and the negative weights represent suppression. (2) An additive sum. It is used to find the sum of the

corresponding synaptic weights of the input signal to the neurons. (3) An activation function is used to limit the output amplitude of neurons. An activation function is also known as a suppression function because it suppresses (limits) the input signal to a certain value within the allowable range. (4) Threshold. Besides, you can add an external bias to a neuron model, which is to increase or decrease the network input of the activation function.it is explain in Fig. (3-5).

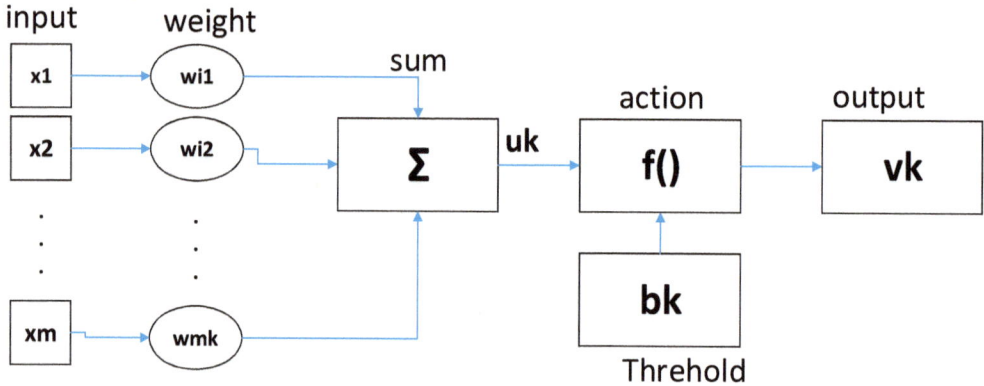

Fig. (3-5). The example of neural networks.

The additive sum is calculated use formula 3-12.

$$u_k = \sum_{i=1}^{m} w_{ik} x_k \qquad \text{(3-12)}$$

Output is calculated use formula 3-13.

$$v_k = f(u_k + b_k) \qquad \text{(3-13)}$$

Activation function normally include:

$$f(u) = \begin{cases} 1, u \geq 0 \\ 0, u \leq 0 \end{cases} \qquad \text{(3-14)}$$

Or

$$f(u) = \begin{cases} 1, u \geq 1 \\ u, -1 \leq u \leq 1 \\ -1, u \leq -1 \end{cases} \qquad \text{(3-15)}$$

Or

$$f(u) = \frac{1}{1+e^{-u}} \qquad\qquad (3\text{-}16)$$

It abstracts the human brain neuron network from the perspective of information processing, establishes a simple model, and forms different networks according to different connection modes. In engineering and academia, it is often referred to as neural network or similar neural network.

Neural network is an operation model, which consists of a large number of nodes (or neurons) connected with each other. Each node represents a specific output function, called activation function. The connection between two nodes represents a weighted value for the signal passing through the connection, which is called the weight, which is equivalent to the memory of the artificial neural network. The output of the network varies according to the connection mode, weight value and excitation function of the network. The network itself is usually an approximation of an algorithm or function in nature, or it may be an expression of a logical strategy. The Fig. (**3-6**) illustrates the multi-input and multi-output in Neural network.

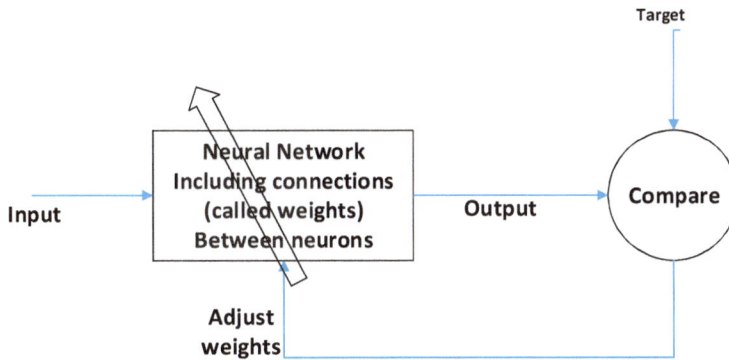

Fig. (3-6). The weights adjust of neural networks.

Artificial Neural Network (ANN) is a parallel distributed system. It adopts a completely different mechanism from traditional AI and information processing technology. It overcomes the shortcomings of traditional logic symbol-based AI in dealing with intuitive and unstructured information. It has the characteristics of self-adaptation, self-organization and real-time learning. The structure of ANN is shown in Fig. (**3-7**).

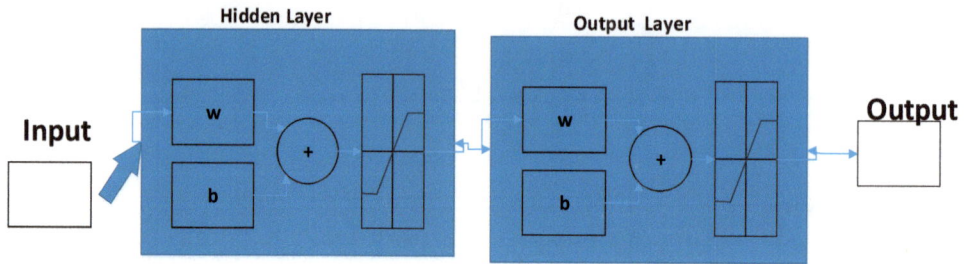

Fig. (3-7). The structure of neural networks.

The BP Neural networks works use the following step:

Step1: initiation

Step2: output calculation of hidden layer

Step 3:output calculation of output layer

Step 4: error calculation, Ok is desire output

$$e_k = O_k - v_k \tag{3-17}$$

Step 5: weights correction. ç is rate of learning.

$$w_{ij} = w_{ij} + \eta u_j(1 - u_j)x(i) \sum_{k=1}^{m} w_{jk}\, e_k \tag{3-18}$$

$$w_{jk} = w_{jk} + \eta H_j e_k \tag{3-19}$$

Step 6:Update threshold

$$a_j = a_i + \eta H_j(1 - H_j) \sum_{k=1}^{m} w_{jk}\, e_k \tag{3-18}$$

$$b_k = b_k + e_k \tag{3-20}$$

Exemple 3-18

SamNum=100; TestSamNum=100;

HiddenUnitNum=10;

InDim=1; OutDim=1;

```
SamIn=0.01*pi:0.02*pi:2*pi;

SamOut=sin(SamIn);

TestSamIn=0.01*pi:0.02*pi:2*pi;

TestSamOut=sin(TestSamIn);

figure

hold on

grid

plot(TestSamIn,TestSamOut,'k--')

xlabel('Input x');

ylabel('Output y');

MaxEpochs=50000;

lr=0.005;

E0=1;

W1=0.2*rand(HiddenUnitNum,InDim)-0.1;

B1=0.2*rand(HiddenUnitNum,1)-0.1;

W2=0.2*rand(OutDim,HiddenUnitNum)-0.1;

B2=0.2*rand(OutDim,1)-0.1;

W1Ex=[W1 B1]

W2Ex=[W2 B2]

SamInEx=[SamIn' ones(SamNum,1)]'

ErrHistory=[];

for i=1:MaxEpochs

HiddenOut=logsig(W1Ex*SamInEx);

HiddenOutEx=[HiddenOut' ones(SamNum,1)]';
```

```
NetworkOut=W2Ex*HiddenOutEx;

Error=SamOut-NetworkOut;

SSE=sumsqr(Error);

ErrHistory=[ErrHistory SSE];

switch round(SSE*10)

case 4

lr=0.003;

case 3

lr=0.001;

case 2

lr=0.0005;

case 1

lr=0.01;

case 0

break;

otherwise

lr=0.005;

end

Delta2=Error;

Delta1=W2'*Delta2.*HiddenOut.*(1-HiddenOut);

dW2Ex=Delta2*HiddenOutEx';

dW1Ex=Delta1*SamInEx';

W1Ex=W1Ex+lr*dW1Ex;

W2Ex=W2Ex+lr*dW2Ex;
```

```
W2=W2Ex(:,1:HiddenUnitNum);

end

i

W1=W1Ex(:,1:InDim)

B1=W1Ex(:,InDim+1)

W2

B2=W2Ex(:,1+HiddenUnitNum);

TestHiddenOut=logsig(W1*TestSamIn+repmat(B1,1,TestSamNum));

TestNNOut=W2*TestHiddenOut+repmat(B2,1,TestSamNum);

plot(TestSamIn,TestNNOut,'r*')

figure

hold on

grid

[xx,Num]=size(ErrHistory);

plot(1:Num,ErrHistory,'k-');
```

3.4.3. The Global Optimization Problem

Genetic Algorithms (GA) is a computational model to simulate the natural selection and genetic mechanism of Darwin's biological evolution theory. It is a method to search the optimal solution by simulating the natural evolution process. Genetic algorithm begins with a population representing the potential solution set of the problem, and a population consists of a certain number of individuals encoded by genes. Each individual is actually a characteristic entity of the chromosome. As the main carrier of genetic material, the chromosome is the collection of multiple genes. Its internal expression (*i.e.* genotype) is a kind of genome combination. It determines the external expression of individual shape. For example, the characteristics of black hair are determined by some kind of genome combination which controls this feature in the chromosome. Therefore, the mapping from phenotype to genotype, *i.e.* coding, is needed at the beginning.

Because the work of copying gene coding is very complex, we often simplify it, such as binary coding. After the generation of the first-generation population, according to the survival of the fittest and the survival of the fittest, generation by generation evolves to produce better and better approximate solutions. In each generation, according to the fitness of the individual in the problem domain. Individuals are selected and crossover and mutation are carried out with the help of genetic operators of natural genetics to generate a population representing a new solution set. This process will result in a population that is more adaptable to the environment than its predecessors, like natural evolution. The decoding of the optimal individuals in the last generation can be used as an approximate optimal solution to the problem [10].

Genetic algorithm is a highly parallel, stochastic and adaptive optimization algorithm based on "survival of the fittest". By replication, crossover, and mutation, the "chromosome" group represented by the problem solution coding evolves from generation to generation, and eventually converges to the most suitable group, to obtain the optimal or satisfactory solution of the problem. Its advantages are simple principle and operation, strong versatility, unconstrained conditions, implicit parallelism and global searchability, which are widely used in combinatorial optimization problems. The earliest application of genetic algorithm to job-shop scheduling problems is that Davis genetic algorithm uses less neighborhood knowledge when solving job-shop scheduling problem, which is more suitable for practical application. How to use genetic algorithm to solve job-shop scheduling problems efficiently has always been considered a challenging problem and become a research hotspot.

The basic operation procedure of genetic algorithm is as follows:

A) Initialization: Set the evolutionary algebra counter $t = 0$, set the maximum evolutionary algebra T, and randomly generate M individuals as the initial population P (0).

B) Individual evaluation: Calculate the fitness of individuals in group P (t).

C) Selection operation: the selection operator is applied to the group. The purpose of selection is to inherit the optimized individuals directly to the next generation or to produce new individuals through paired crossover and then to the next generation. Selection is based on the fitness evaluation of individuals in a group.

D) Crossover operation: the crossover operator is applied to the population. The crossover operator plays a key role in genetic algorithm.

E) Mutation operation: the mutation operator is applied to the population. That is, to change the gene value at some loci of individual strands in a population. Population P (t) is selected, crossed and mutated to obtain the next generation population P (t + 1).

F) Termination condition judgment: if t = T, the maximum fitness individuals obtained in the evolutionary process are taken as the output of the optimal solution to terminate the calculation.

Genetic algorithm is a general algorithm to solve the search problem. It can be used for all kinds of common problems. The common features of the search algorithm are:

1) Firstly, a group of candidate solutions is formed.

2) Calculating the fitness of these candidate solutions according to some adaptive conditions

3) Retain some candidate solutions according to fitness and abandon others.

4) Some operations are performed on the reserved candidate solutions to generate new candidate solutions.

In genetic algorithm, these features are combined in a special way: parallel search based on chromosome group, selection operation with guessing property, exchange operation and mutation operation. This special combination distinguishes genetic algorithm from other search algorithms

(1) Code

Genetic algorithms cannot directly deal with the parameters of the problem space. They must be transformed into chromosomes of individuals composed of genes according to a certain structure in the genetic space. This transformation operation is called encoding, or representation.

The following three criteria are oftcn used in evaluating coding strategies:

A) Completeness: All points (candidate solutions) in the problem space can be represented as points (chromosomes) in the GA space.

B) Soundness: chromosomes in GA space can correspond to candidate solutions in all problem spaces.

C) Nonredundancy: chromosomes correspond to candidate solutions one by one.

At present, several commonly used encoding technologies are binary encoding, floating-point encoding, character encoding, encoding, *etc.*

Binary coding is the most commonly used coding method in genetic algorithm. That is, the binary character set {0,1} generates the usual 0,1 string to represent the candidate solution of the problem space. It has the following characteristics:

A) Simple and easy

B) Compliance with the Minimum Character Set Coding Principle.

C) It is easy to use the mode theorem for analysis because the mode theorem is based on it.

(2) Fitness function

In evolutionary theory, fitness is the ability of an individual to adapt to the environment and to reproduce its offspring. The fitness function of genetic algorithm is also called evaluation function. It is used to judge the quality of individuals in a group. It is evaluated according to the objective function of the problem.

Generally, genetic algorithm does not need other external information in the process of search and evolution, only uses evaluation function to evaluate the advantages and disadvantages of individuals or solutions, and serves as the basis for future genetic operations. In genetic algorithm, the fitness function is ranked by comparison and the selection probability is calculated on this basis, so the fitness function value should be positive. Thus, in many cases, it is necessary to map the objective function to the fitness function which is in the form of maximum value and whose value is not negative.

The design of fitness function mainly satisfies the following conditions:

A) Single-valued, continuous, non-negative and maximized

B) Reasonable and consistent

C) Small amount of calculation

D) Strong versatility.

In the specific application, the design of fitness function depends on the requirement of solving the problem itself. The design of fitness function directly affects the performance of genetic algorithm.

(3) Initial group selection

Individuals in the initial population of genetic algorithms are randomly generated. Generally speaking, the following strategies can be adopted to set up the initial group:

(a) According to the inherent knowledge of the problem, try to grasp the distribution range of the space occupied by the optimal solution in the whole problem space, and then set the initial population within the distribution range.

B) A certain number of individuals are randomly generated, and then the best individuals are selected and added to the initial population. This process continues to iterate until the number of individuals in the initial population reaches a predetermined scale

Example 3-19:

```
%% Start of Program

clc

clear

tic

%% Algorithm Parameters

SelMethod = 1;

CrossMethod = 1;

PopSize = 100;

MaxIteration = 1000;

CrossPercent = 70;

MutatPercent = 20;

ElitPercent = 100 - CrossPercent - MutatPercent;

CrossNum = round(CrossPercent/100*PopSize);

if mod(CrossNum,2)~=0;

CrossNum = CrossNum - 1;

end
```

```
MutatNum = round(MutatPercent/100*PopSize);

ElitNum = PopSize - CrossNum - MutatNum;

%% Problem Satement

VarMin = -100;

VarMax =100;

DimNum = 30;

CostFuncName =@testfunc3;

%% Initial Population

Pop = rand(PopSize,DimNum) * (VarMax - VarMin) + VarMin;

Cost = feval(CostFuncName,Pop);

[Cost Indx] = sort(Cost);

Pop = Pop(Indx,:);

%% Main Loop

MeanMat = [];

MinMat = [];

for Iter = 1:MaxIteration

%% Elitism

ElitPop = Pop(1:ElitNum,:);


%% Cross Over

CrossPop = [];

ParentIndexes = SelectParents_Fcn(Cost,CrossNum,SelMethod);


for ii = 1:CrossNum/2

Par1Indx = ParentIndexes(ii*2-1);
```

```
Par2Indx = ParentIndexes(ii*2);

Par1 = Pop(Par1Indx,:);

Par2 = Pop(Par2Indx,:);

[Off1 Off2] = MyCrossOver_Fcn(Par1,Par2,CrossMethod);

CrossPop = [CrossPop ; Off1 ; Off2];

end

%% Mutation

MutatPop = rand(MutatNum,DimNum) * (VarMax - VarMin) + VarMin;

%% New Population

Pop = [ElitPop ; CrossPop ; MutatPop];

Cost = feval(CostFuncName,Pop);

[Cost Indx] = sort(Cost);

Pop = Pop(Indx,:);

%% Algorithm Progress

disp('----------------------------------------------')

BestP = Pop(1,:)

BestC = Cost(1)

MinMat(Iter) = Cost(1);

MeanMat(Iter) = mean(Cost);

%plot(MinMat,'--r','linewidth',2);

%hold on

% plot(MeanMat,'--k','linewidth',2);
```

```
% hold off

%pause(.5)

semilogy(Iter,MinMat(Iter),'r.')

hold on

semilogy(Iter,MeanMat(Iter),'b.')

end

% ylim([0 5])

%%% Results

BestSolution = Pop(1,:)

BestCost = Cost(1,:)

function [Off1 Off2] = MyCrossOver_Fcn(Par1,Par2,CrossMethod)

switch CrossMethod

case 1

Beta1 = rand;

Beta2 = rand;

Off1 = Beta1*Par1 + (1-Beta1)*Par2;

Off2 = Beta2*Par1 + (1-Beta2)*Par2;

case 2

case 3

end

%%% End of Program

toc

function Cost = MyFucn_Fcn(Pop)
```

```
Cost = zeros(size(Pop,1),1);

for ii = 1:size(Pop,1)

p = Pop(ii,:);

C = sum(p.^2);

Cost(ii,1) = C;

end

end

function ParIndexes = SelectParents_Fcn(Cost,SelectionNum,SelMethod)

PopSize = size(Cost,1);

switch SelMethod

case 1

R = randperm(PopSize); ParIndexes = R(1:SelectionNum);

end

end
```

Locatelli, Marco, and F. Schoen. Global optimization. Theory, algorithms, and applications. Global optimization: theory, algorithms, and applications. 2013.

PROBLEM

3-1 What are the commonly used digital filtering algorithms? Explain the characteristics and application occasions of various filtering algorithms.

3-2 What are the causes of zero error? What are the reasons for the gain error? The correction method is described briefly.

3-3 A set of discrete data reflecting the measured value is obtained through measurement. To establish an approximate mathematical model reflecting the change of the measured value, what are the commonly used modeling methods?

3-4 Illustrate the concept of scaling transformation with examples

3-5 What is computational intelligence? This paper briefly describes the application of Computational Intelligence in data analysis and processing and its algorithm.

3-6 Write a program with MATLAB Wavelet Tool Wave to filter a noisy signal?

3-7 Write a neural network optimization algorithm with MATLAB neural network toolkit to optimize a set of data.

3-8 An interface circuit of programmable gain amplifier controlled by 8031 single chip computers is designed. It is known that the input signal is less than 10 mV. When the input signal is less than 1 mV, the gain is 1000. When the input signal is increased by 1 mV, the gain is automatically doubled until 100 mV.

3-9 The sampling object of a data acquisition system is the temperature and humidity of the greenhouse, which requires the relative errors of positive and negative 1% and positive and negative 3% respectively. When collecting data every 10 minutes, what type of A/D converter and channel scheme should be selected?

3-10 A data acquisition system with a sampling/holding device, whose sampling frequency is 100kHz, FSR = 10V, aperture time is 3ns, n = 8. Is the sampling frequency too high?

REFERENCES

[1]   "Uncertainty" available from https://en.wikipedia.org/wiki/ [Accessed: 2019.2.1].

[2]   D.F. Chen, and Q. Lin, *The Intelligent Instrument*. The China Machine Press: Beijing, China, 2014.

[3]   V. Deshpande, "Development and simulation of SPWM and SVPWM control induction motor drive." International Conference on Emerging Trends in Engineering & Technology 2009.

[4]   A. O'Dwyer, "Handbook of PI and PID Controller Tuning Rules", *Automatica,* vol. 41, no. 2, pp. 355-356, 2006. [analog-to-digital converter" available from https://en.wikipedia.org/wiki/ analog-to-digital converter.].

[5]   L. Fu, and C.J. Deng, Theory, design and application of intelligent instruments. China, Chengdu: Southwest Jiao tong University Press, 2014.

[6]   "Inverse problem" available from https://en.wikipedia.org/wiki/ [Accessed: 2019.2.1].

[7]   "Regularization" available from https://en.wikipedia.org/wiki/ [Accessed: 2019.2.1].

[8]   "Deep learning" available from https://en.wikipedia.org/wiki/ [Accessed: 2019.2.1].

[9]   "Neural networks" https://en.wikipedia.org/wiki/ [Accessed: 2019.2.1].

[10] Locatelli, Marco, and F. Schoen. Global optimization. Theory, algorithms, and applications. Global optimization: theory, algorithms, and applications. 2013.

## CHAPTER 4

The Performance Analysis of Intelligent Instrument

Abstract: In this chapter, we will discuss the accuracy, reaction speed, and their design theory; the software and hardware tests of intelligent instrument; and the reliability engineering of intelligent instrument. Some context includes the design of experiments, the quality of data, the user-centered design and so on.

Keywords: Accuracy, Design of experiments, FTA, Intelligent instrument design theory, Quality, Reliability, Software test, SCA, User-centered design.

4.1. THE ACCURACY AND INTELLIGENT INSTRUMENT DESIGN THEORY

The precision design of instrument is the inverse problem of instrument precision synthesis, and its basic task is to allocate the total error of the given instrument to each component of the instrument reasonably, and to provide the basis for the correct design of each component structure of the instrument and to make the tolerance and technical requirements of the parts [1].

4.1.1. The Accuracy Design Theory

(1) The principle of Micro Error [2]:

A trivial error is called a small error if the effect of this error contributes a small factor of the total effect, (*e.g.* the effect value of this error is less than 1/10 of the total effect value). When the measurement process contains a variety of errors, when an error almost not affect the total error of the measurement result, the error can be ignored. For example, the standard error of the system is δ_x, it is formed by its components δ_x; $i \in 1, n$:

$$\sigma_x = \sqrt{\sigma_1^2 + \sigma_2^2 + \cdots + \sigma_n^2} \qquad (4\text{-}1)$$

The principle of Micro Error shows: if one of its component's errors e_k was ignored, it should follow $e_k < \frac{1}{10} e_x$. In this case, the standard error δ_y is equal to δ_x in its accuracy range.

$$\sigma_y = \sqrt{\sigma_1^2 + \cdots + \sigma_{k-1}{}^2 + \sigma_{k+1}{}^2 + \cdots + \sigma_n^2} \qquad (4\text{-}2)$$

This principle can also be described as uncertainty. In this way, it is called as 1/3 principle. The uncertainty of the system is U. It follows the formula (4-3).

$$U = \sqrt{U_1^2 + U_2^2} \qquad (4\text{-}3)$$

As the principle of Micro Error, if U_2 can be ignored.

$$\sqrt{U_1^2 + U_2^2} - U_2 \leq \frac{1}{10}\sqrt{U_1^2 + U_2^2} \qquad (4\text{-}4)$$

Then the $U_2 \leq \frac{U}{3}$, so it is also called as 1/3 principle

(2) MCP Detection capability Index method [2]

The measuring instrument is divided into three categories according to the nature of the measurement:

Parameter checking: The range of parameters is usually judged.

Parameter monitoring: Between parameter inspection and measurement, the parameters of measurement and control.

Parameter measurement: usually refers to the exact value or measurement value of a parameter.

For precision differentiation, the detection capability index is defined using formula(4-5):

$$M_{cp} = \frac{T}{6u} = \frac{T}{2U} \qquad (4\text{-}5)$$

The u is the standard uncertainty of the measurement result; U is the total uncertainty of the measurement result, T is the range of variation allowed by the parameter, and when the measurement error is estimated, the U_1 is the uncertainty of the measuring instrument.

$$M_{cp} = \frac{2\Delta_{permit}}{3U_1}$$ (4-6)

Example 4-1:

A measuring instrument is designed to test the port metering import and export bulk grain, and its detection condition is required to be A grade, and the accuracy of the measuring instrument is determined.

Solution: According to international practice, the error range of the port metering grain is ±0.4%, otherwise it is compensated, so the measurement error of the object under test is ±0.4%, for A-level measurement, check the detection Capability Index Table, M_{cp}, then

$$M_{cp} = \frac{2\Delta_{permit}}{3U_1}$$ (4-7)

$$U_1 = \frac{2\Delta_{permit}}{3M_{cp}} = \frac{2 \times 0.4\%}{3 \times (1.7 \sim 2)} = (0.15\% \sim 0.13\%)$$ (4-8)

(3) The empirical principles of measurement & control instrument design

A. The principle of Abéché and its extension

B. Deformation minimum and the measures to reduce the influence of deformation

C. Minimum principle of measurement chain

D. The uniform principle of coordinate system datum

E. Precision Matching principle

F. Economic Principles

1) Abéché and its extension:

or the gauge to give the correct measurement results, the reading line of the instrument must be placed on the extension line of the measured dimension.

Example 4-2:

Vernier caliper that does not use the Abéché principle, and one is a length gauge using the Abéché principle.

Formula (4-9) is the error Δ_1 analysis of the vernier caliper, the corner of which is ϕ, and the distance is s.

$$\Delta_1 = s \cdot tan\varphi \tag{4-9}$$
$$Ex: s = 20mm, \varphi = 1' \Longrightarrow \Delta_1 = 20 \cdot 0.0003 = 0.006mm$$

The formula 4-10 is the error analysis of the Abéché aperture, and its deflection angle is ϕ and the length is d.

$$\tag{4-10}$$
$$\Delta_1 = d - d' = d(1 - cos\varphi) \cong \frac{d \cdot \varphi^2}{2}$$
$$x: s = 20mm, \varphi = 1' \Longrightarrow \Delta_1 = \frac{20 \cdot 0.0003^2}{2} = 9 \times 10^{-7} mm$$

2) The principle of minimum deformation and the measures to reduce the influence of deformation

In the process of instrument operation, the instrument structure or instrument parameters varying with the force or the temperature should be avoided, and their influence on the accuracy of the instrument should be minimized.

To design a sensor, it is a principle that the materials with less temperature and humidity changes are more suitable to be selected as capacitance or inductive sensor.

Example 4-3: Strain measurement bridge zero-point temperature drift compensation.

Solution:

A, balance conditions: There are two equilibrium conditions for the bridge: One is the equilibrium condition of the resistor R1*R4=R2*R3, the other is that the temperature coefficient has a a1+a4=a2+a3.

B, strain measurement Bridge zero-point temperature drift circuit

The zero-point temperature drift is caused by the inconsistency of the resistance values of four diffusion resistors and their temperature coefficients. Generally, the series, parallel resistance compensation method is used, R_s is a series resistance; R_p is a parallel resistor. The series resistance mainly plays the role of zero correction, and the parallel resistance mainly plays a temperature compensating role. As is shown in Fig. (**4-1**).

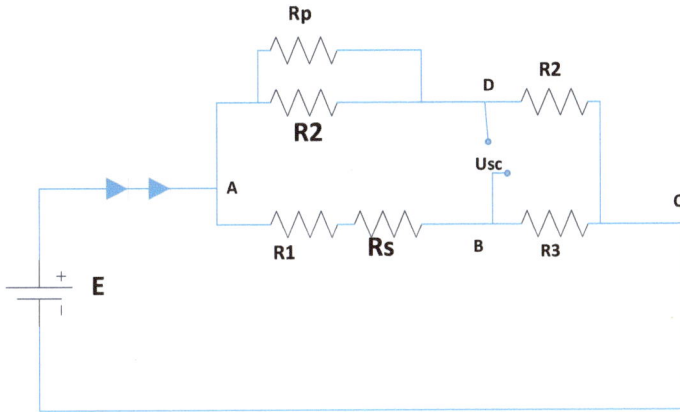

Fig. (4-1). The strain measurement bridge.

3) Minimum principle of measuring chain

The number of components that make up the instrument's measurement chain should be minimal. All relevant components include the measured parts, the standard parts, the sensory elements, the positioning elements. They belong to the measurement chain. Such as electronic displacement synchronous comparison principle.

4) Uniform principle of coordinate system datum

The position of the instrument group is unified (for example, the design base of the part, the process base and the measurement base are consistent, the transformation relationship between the instrument sub-coordinate system and the main coordinate systems is the linear function).

5) Accuracy matching principle

On the basis of the precision analysis of the instrument, according to the difference of the influence degree of each part of the instrument on the precision of the instrument, the precision requirements and the appropriate precision distribution of each part are suitable respectively.

Example 4-4: An Intelligent Sensing Instrumentation System [3]

Solution:

The lower and middle level structure of ISIS measurement channel is shown in Fig. (**4-2**) [3].

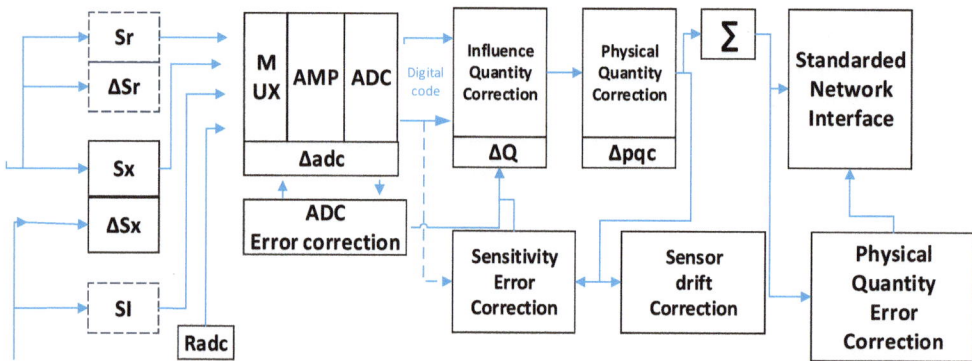

Fig. (4-2). General structure of ISIS measurement channel (lower and middle levels) [3].

The three levels of ISIS includes:

• The lower level: Sensor, Sensor correction, ADC, ADC error correction.

• The middle level: Influence Quantity correction, and Physical Quantity correction; include sensitivity error correction, sensor drift correction; most include a standard network interface.

• The higher level: Call the function to control the self-modification, include processing and storage.

4.1.2. The Design of Experiments

It is a branch of mathematical statistics. The mathematical principle and implementation method of how to formulate an appropriate experimental scheme in accordance with the predetermined objectives in order to facilitate the effective statistical analysis of experimental results.

The design of an experiment, that is, an arrangement of the experiment need to consider the type of problems, the degree of universality, the degree of effectiveness, the homogeneity of the experimental units, the cost and time consuming of each experiment, and so on. The appropriate factors and corresponding levels should be selected to give the facts. Specific procedures for testing implementation and framework for data analysis.

Product quality is mainly determined by design. A good test design includes several aspects.

The first is to measure the quality of the products. The management emphasizes the use of data. Therefore, this quality index must be a quantifiable index, which is called test index in test design, also known as response variable or output variable.

The second is to find out the possible factors that affect the test indicators, also known as impact factors and input variables. Various state of factor change is called the level, which requires that the scope of factor level be determined preliminarily according to professional knowledge.

Thirdly, according to the actual problems, the suitable design method of experiment is selected. There are many methods of experimental design, each method has different applicable conditions. Choosing the suitable method can achieve twice the result with half the effort. If the method is not correct or effective, experimental design cannot be carried out at all, it will achieve twice the result with half the effort.

The fourth is to scientifically analyze the test results, including the direct analysis of data, variance analysis, regression analysis and other statistical analysis methods, which can be completed with the help of Minitab software.

The three basic principles of experimental design are repetition, randomization and block.

Repetition means the repetition of basic experiments. Repetition has two important properties. First, the experimenter is allowed to obtain an estimate of the test error. The estimator of this error becomes the basic unit of measurement to determine whether the observed diffcrence of the data is a statistical test difference. Second, if the sample means is used as an estimate of the effect of a factor in the experiment, repetition allows the experimenter to obtain a more accurate estimate of the effect. If S2 is the variance of data and N repetitions occur, the variance of the sample mean is s2/n. The practical meaning of this point is that if $n = 1$, if $y1 = 145$, and $y2 = 147$, then we may be able to deduce whether there is any difference between the two treatments, that is to say, observation difference $147-145 = 2$ may be the result of experimental error. However, if n is reasonable enough and the test error is small enough, we can see that y1 randomization is the cornerstone of the statistical method used in experimental design.

The so-called randomization refers to the distribution of test materials and the order of each test, which is determined randomly. Statistical methods require observations (or errors) to be independent random variables. Randomization usually makes this assumption valid. Appropriate randomization of the experiment also contributes to the "uniform" effect of possible external factors.

Blocking is a method to improve the accuracy of the test. A block is a part of the test material, and its properties should be more similar than those of all the test materials themselves. Blocking involves comparing experimental conditions of interest within each block.

The experimental design methods can be divided into two categories, one is the orthogonal experimental design method, the other is factorial method.

Category I: Orthogonal experimental design method

• Definition: Orthogonal experimental design is a scientific method to study and deal with multifactor experiments. It uses a standardized form, orthogonal table, to select test conditions, arrange test plans and carry out tests, and through fewer tests, find out better production conditions, that is, the best or better test scheme.

• Use Orthogonal experimental design is mainly used to investigate some characteristics of complex systems (products, processes) or the effects of multiple factors on certain characteristics of systems (products, processes), to identify the more influential factors in the system, the magnitude of their impacts, and the possible interrelationships among them, to promote product design and development and process. Optimize, control or improve existing products (or systems).

• Form of tables

Category II: Factorial method

• Definition Analysis: Factorial test design, factorial test and so on. It is an effective method to study the effects of two or more changing factors. Many experiments require examining the effects of two or more variables. For example, there are several factors: the effect on product quality; the effect on a certain machine; the effect on the performance of a certain material; the effect on the combustion consumption of a certain process and so on. The factors studied are tested successively according to all levels (ranks) of all factors, which is called factorial test, or complete factorial test, abbreviated as factorial method.

• Use for new product development, product or process improvement, as well as installation services, through fewer tests, find a combination of high-quality, high-yield, low-consumption factors, to achieve the purpose of improvement.

• Form of tables

Example 4-5: Improve an Engine Cooling Fan Using Design for Six Sigma Techniques

Solution:

Using a Design for Six Sigma approach: Define, Measure, Analyze, Improve, and Control (DMAIC).

• Define the Problem

• Assess Cooling Fan Performance

• Determine Factors That Affect Fan Performance

• Improve the Cooling Fan Performance

• Sensitivity Analysis

• Control Manufacturing of the Improved Cooling Fan

cd(matlabroot)

cd('help/toolbox/stats/examples')

load OriginalFan

plot(originalfan)

xlabel('Observation')

ylabel('Max Airflow (ft^3/min)')

title('Historical Production Data')

figure()

histfit(originalfan) % Plot histogram with normal distribution fit

format shortg

xlabel('Airflow (ft^3/min)')

```
ylabel('Frequency (counts)')

title('Airflow Histogram')

pd = fitdist(originalfan,'normal') % Fit normal distribution to data

CodedValue = bbdesign(3)

runorder = randperm(15); % Random permutation of the runs

bounds = [1 1.5;15 35;1 2]; % Min and max values for each factor

RealValue = zeros(size(CodedValue));

for i = 1:size(CodedValue,2) % Convert coded values to real-world units

zmax = max(CodedValue(:,i));

zmin = min(CodedValue(:,i));

RealValue(:,i) = interp1([zmin zmax],bounds(i,:),CodedValue(:,i));

end

TestResult = [837 864 829 856 880 879 872 874 834 833 860 859 874 876 875]';

disp({'Run Number','Distance','Pitch','Clearance','Airflow'})

disp(sortrows([runorder' RealValue TestResult]))

Expmt = table(runorder', CodedValue(:,1), CodedValue(:,2), CodedValue(:,3), ...

TestResult,'VariableNames',{'RunNumber','D','P','C','Airflow'});

mdl = fitlm(Expmt,'Airflow~D*P*C-D:P:C+D^2+P^2+C^2');

figure()

h = bar(mdl.Coefficients.Estimate(2:10));

set(h,'facecolor',[0.8 0.8 0.9])

legend('Coefficient')

set(gcf,'units','normalized','position',[0.05 0.4 0.35 0.4])

set(gca,'xticklabel',mdl.CoefficientNames(2:10))
```

```
ylabel('Airflow (ft^3/min)')

xlabel('Normalized Coefficient')

title('Quadratic Model Coefficients')

plotSlice(mdl)

f = @(x) -x2fx(x,'quadratic')*mdl.Coefficients.Estimate;

lb = [-1 -1 -1]; % Lower bound

ub = [1, 1, 1]; % Upper bound

x0 = [0 0 0]; % Starting point

[optfactors,fval] = fmincon(f,x0,[],[],[],[],lb,ub,[]); % Invoke the solver

maxval = -fval;

maxloc = (optfactors + 1)';

bounds = [1 1.5;15 35;1 2];

maxloc=bounds(:,1)+maxloc .* ((bounds(:,2) - bounds(:,1))/2);

disp('Optimal Values:')

disp({'Distance','Pitch','Clearance','Airflow'})

disp([maxloc' maxval])

load AirflowData

tbl = table(pitch,airflow);

mdl2 = fitlm(tbl,'airflow~pitch^2');

mdl2.Rsquared.Ordinary

figure()

plot(pitch,airflow,'.r')

hold on

ylim([840 885])
```

```
line(pitch,mdl2.Fitted,'color','b')

title('Fitted Model and Data')

xlabel('Pitch angle (degrees)')

ylabel('Airflow (ft^3/min)')

legend('Test data','Quadratic model','Location','se')

hold off

pitch(find(airflow==max(airflow)))

rng('default')

dist = random('normal',optfactors(1),0.20,[10000 1]);

pitch = random('normal',optfactors(2),0.028,[10000 1]);

clearance = random('normal',optfactors(3),0.25,[10000 1]);

noise = random('normal',0,mdl2.RMSE,[10000 1]);

simfactor = [dist pitch clearance];

X = x2fx(simfactor,'quadratic');

simflow = X*mdl.Coefficients.Estimate+noise;

pd = fitdist(simflow,'normal');

histfit(simflow)

hold on

text(pd.mu+2,300,['Mean: ' num2str(round(pd.mu))])

text(pd.mu+2,280,['Standard deviation: ' num2str(round(pd.sigma))])

hold off

xlabel('Airflow (ft^3/min)')

ylabel('Frequency')

title('Monte Carlo Simulation Results')
```

```
format long

pfail = cdf(pd,875)

pass = (1-pfail)*100

S = capability(simflow,[875.0 890])

pass = (1-S.Pl)*100

load spcdata

figure()

controlchart(spcflow,'chart',{'xbar','s'}) % Reshape the data into daily sets

xlabel('Day')

[row,col] = size(spcflow);

S2 = capability(reshape(spcflow,row*col,1),[875.0 890])

pass = (1-S.Pl)*100
```

Example 4-6:

1) Accuracy can be defined using true value and precision. in industrial instrumentation. It includes measurement tolerance.

2) Accuracy can be defined using trueness (proximity of measurement results to the true value) and precision (repeatability or reproducibility of the measurement) [5].

4.2. SOFTWARE TEST

It describes a process used to promote the validity, integrity, security and quality of authentication software. In other words, software testing is a process of auditing or comparing actual output with the expected output. The classical definition of software testing is the process of operating a program under specified conditions to detect program errors, measure software quality, and evaluate whether it meets design requirements [6].

Software testing is defined in the software engineering terminology proposed by IEEE as "the process of running or measuring a software system by manual or automatic means, whose purpose is to verify whether it meets the specified requirements or to clarify the difference between the expected results and the actual results". This definition clearly states that the purpose of software testing is to test whether the software system meets the requirements. It is no longer a one-time activity, but only activity in the later stage of development. It is integrated with the whole development process. Software testing has become a profession, which requires the use of specialized methods and means, and requires specialized personnel and experts to undertake.

1). Static Test Method

Static testing refers to the static analysis test of software code. There is fewer data applied in this kind of process. The main process is to pass the static testing of software. (*i.e.* manual inference or computer-aided testing) test the correctness of arithmetic and arithmetic in the program, and then complete the testing process. The advantage of this kind of testing is that it can consume less time and resources to complete the testing of software and software code, and can find the errors in this kind of code more clearly. Static testing method has a wide range of applications, especially for larger software testing.

2). Dynamic testing

The main purpose of computer dynamic testing is to detect the problems in software operation. Compared with static testing, the dynamic testing method is called dynamic because it mainly depends on the application of programs. The main purpose is to detect whether the dynamic behavior in software is missing and whether the software is running well. The most obvious feature is that the software is running when dynamic testing is carried out. Only in this way can software defects be found in the process of use, and then such defects can be repaired. At present, the dynamic testing process can include two kinds of factors, namely, the software under test and the data needed in the test. The two kinds of factors determine the correct and effective development of dynamic testing.

3). White Box Testing

White-box testing is transparent relative to black-box testing. The principle of white-box testing is to debug the internal working process of the product according to the internal application and source code of the software. In the process of testing, it is often analyzed in coordination with the internal structure of the software. The greatest advantage is that it can effectively solve the problems arising in the internal application program of the software. In the process of testing, it is often combined with black-box testing. When testing software has more functions, the white-box testing method can also debug the situation effectively. Among them, decision testing is one of the most important test program structures in white-box testing method. As the overall realization of program logic structure, this kind of program structure plays an important role in program testing. This kind of testing method covers a wide range of code types in the program and is suitable for multi-type programs. In practical testing, white-box testing method is often used with black-box testing method. Taking unknown errors detected in dynamic testing method as an example, first of all, black-box testing method is used. If the input data of the program is the same as the output data, it proves that there is no problem with the internal data. It should be analyzed from the code aspect, and if there is a problem, white-box testing method should be used. Testing method is to analyze the internal structure of the software until the problem is detected and modified in time [7].

And a program with 100 lines of total code is run once with test cases and 75 lines of code are executed. Then, the code coverage = 75%. Logical Coverage mainly includes six Common Coverage Methods: Statement Overlay; Decision Covering (also known as Branch Covering); Conditional Covering (also called Address Covering); Decision/conditional coverage; Conditional Combination Covering; Path coverage.

Figs. (**4-3** and **4-4**) are two example test code for white block testing.

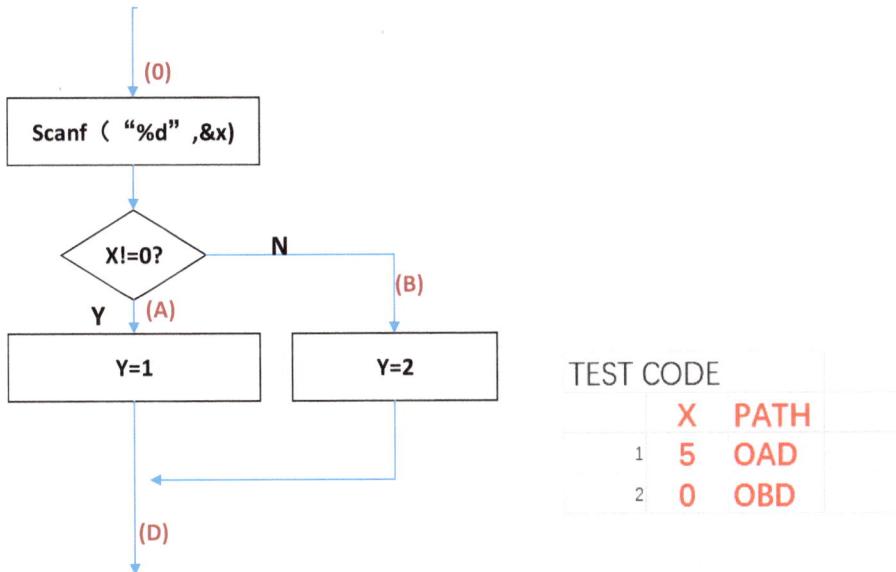

Fig. (**4-3**). One example test code for white block testing.

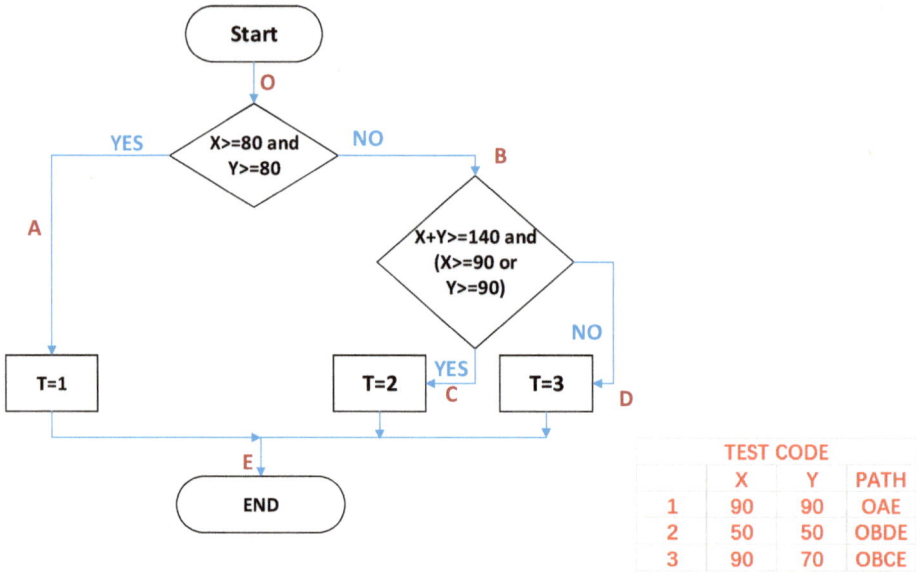

Fig. (4-4). Test code design for white block testing.

	TEST CODE		
	X	Y	PATH
1	90	90	OAE
2	50	50	OBDE
3	90	70	OBCE

4). Black Box Testing

Black box testing, as its name implies, is to simulate the software testing environment as an invisible "black box". Data input and output are observed to check whether the internal functions of the software are normal. When the test is launched, the data is input into the software, waiting for the data output. If the data output is consistent with the predicted data, it will prove that the software passed the test. If the data is different from the predicted data, even if the discrepancy is small, it will also prove that there are problems in the software program, which need to be solved as soon as possible.

Black-box testing methods include equivalence partitioning, boundary value analysis, all-pairs testing, state transition tables, decision table testing, fuzz testing, model-based testing, use case testing, exploratory testing, and specification-based testing.

Example 4-7:

First, the equivalent class division

Q: A program stipulates: "Input three integers a, b, c as the edge length of the three sides to form a triangle." The type of triangle formed by the program determination, when this triangle is a general triangle, isosceles triangle and equilateral triangle, is calculated separately ... "

The equivalent class partitioning method is used to design the test case for the program. (The complexity of the triangle problem is that the relationship between input and output is more complex.)

Solution: Analyze the requirements for input conditions given and implied in the topic: (1) Integer (2) three digits (3) non-zero (4) positive number (5) the sum of the sides is greater than the third side (6) isosceles (7) equilateral.

If a, b, c meet the condition (1) ~ (4), then output one of the following four cases: 1) if the condition (5) is not met, the program output is "non-triangular". 2) If three edges are equal to meet the condition (7), the program output is "equilateral triangle". 3) If only two edges are equal, that is, the condition (6) is met, the program output is "isosceles triangle". 4) If three edges are not equal, the program output is "General triangle". List and Numbering equivalent classes are shown in Table **4-1**.

Test cases that override valid equivalence classes:

a b c overrides equivalent class numbers

3 4 5 (1)-(7)

4 4 5 (1)-(7), (8)

4 5 5 (1)-(7), (9)

5 4 5 (1)-(7), (10)

4 4 4 (1)-(7), (11)

4.3. RELIABILITY ENGINEERING OF INTELLIGENT INSTRUMENT

Reliability in statistics and psychometrics is the overall consistency of measurement [8].

Reliability design: According to the United States Bell Laboratory and Marine Laboratory statistics, due to design problems caused by product failure accounted for more than 40% of the total failure, if you consider the design of the relevant issues, it is about 57%.

Table 4-1. An invalid equivalent class.

input	Input	equivalence class	Num	No valid equivalence class	Number
	Three integer number	integer	1	a is not an integer	12
				b is not an integer	13
				c is not an integer	14
				a,b is not an integer	15
				a,c is not an integer	16
				b,c are not an integer	17
				a,b,c are not integer	18
		Three Number	2	Only a	19
				Only b	20
				Only c	21
				Only a,b	22
				Only a,c	23
				Only b,c	24
				Large than a,b,c	25
		No zero number	3	a=0	26
				b=0	27
				c=0	28
				a=b=0	29
				a=c=0	30
				b=c=0	31
				a=b=c=0	32
		Positive number	4	a<0	33
				b<0	34
				c<0	35
				a,b<0	36
				a,c<0	37
				b,c<0	38
				a,b,c<0	39

(Table 4-1) cont.....

output	General triangle	a+b>c	5	a+b=0	40
		a+c>b	6	a+b<0	41
		c+b>a	7	a+c=0	42
				a+c<0	43
				c+b=0	44
				c+b<0	45
	Isosceles triangle	a=b	8		
		a=c	9		
		b=c	10		
	Equilateral triangle	a=b=c	11		

In 1939, the American Aviation Commission compiled the Airworthiness Statistical Notes, which for the first time put forward the aircraft failure rate <or=0.00001 times/h, equivalent to the reliability of the aircraft in an hour.

At the end of World War II, Lussen, a German rocket expert, regarded the V-II rocket induction device as a series system with a reliability of 75%.

The United States learned from the Korean War and established AGREE to solve equipment failure [9].

The reliability design: Now, three types of components (reduction, redundancy, tolerance or fault tolerance) are used to design and manufacture a whole machine. Triple design (system design, parameter design, tolerance design, drift design).

4.3.1. The Concept of Reliability

1) Reliability Definition

Reliability refers to the ability of a product to perform specified functions within a given time interval and under given conditions.

Generalized reliability is the ability of a product to perform specified functions throughout its life cycle, including narrow reliability and maintainability.

The narrow sense of reliability is the failure degree of a product within a specified period.

The probability of reliability is the measured value of reliability. It is the probability function under specified time and conditions.

Reliability is divided into inherent reliability and operational reliability. The former describes the reliability level of product design and manufacturing, and the use of reliability considers the factors of product design, manufacturing, installation environment, maintenance strategy and repair.

Uncertainty refers to the probability that product life T_1 does not exceed a specified time T.

Basic reliability is expressed by mean fault interval time (MTBF), which is expressed by task reliability (MR) and fatal fault interval time (MTBCF).

2) Availability

When the required external resources are satisfied, the product can complete its required functions at a given time or time under given conditions.

3) Maintainability

Under given conditions, the product's ability to maintain or restore the required functional state under given conditions while performing maintenance using the procedures and resources described above.

4) Security

The ability not to cause casualties, harm health, and the environment, or damage or damage to equipment or property.

5) Failure event

The interruption of the product's ability to perform required functions. It refers to the event that the product terminates its ability to ultimately complete the specified function.

6) Failure

Failure refers to the state in which the product cannot perform the specified function.

For a given function, a state of the faulty product is called a failure mode. The intrinsic cause of the faults is the mechanism of the faults.

7) Life expectancy

Duration of product use.

8) Credibility

Credibility is a collective term describing usability and its influencing factors.

9) Defects

Failure to meet requirements related to intended or specified uses [9].

In possibility theory [10]:

Reliability rate (R(t)) is the success rate of the instrument to complete the specified tasks (S(t) is its numbers) under the specified conditions and within the specified time. The total number is N.

$$R(t) = \frac{S(t)}{N} \tag{4-11}$$

Failure rate, also known as instantaneous failure rate or failure rate, refers to the ratio of the number of instruments that fail in a unit time after the instrument runs to T-Time to the number of instruments that are in good condition at t-time.

$$\lambda(t) = \frac{N[R(t) - R(t+\Delta t)]}{NR(t) \cdot \Delta t} \tag{4-12}$$

Random variables consist of continuous and discrete random variable.

Probability distribution: The life cycle distributions function of most systems is often the exponential distribution. If the failure probability of a system is p (it is very small), the probability of non-failure is 1-p=q. Assuming that np is large, x is an event where the system fails. The probability of system failure can be calculated by Poisson distribution.

$$f(x) = \frac{\exp(-\mu) \cdot \mu^x}{x!} = \frac{\exp(-np)(np)^x}{x!} \tag{4-13}$$

In the given timing t, the average of the failure system is $\lambda(t) = np$.

$$f(x, \lambda, t) = \frac{\exp(-\lambda t)(\lambda t)^x}{x!} \tag{4-14}$$

If no one was a failure in this system, then

$$f(0) = \exp(-\lambda t) \tag{4-15}$$

Then

$$R(t) = 1 - F(t) = \int_t^\infty f(t)\, dt \tag{4-16}$$

Example 4-8:

Samples of 2,500 integrated circuits work during the 1000h period. An integrated circuit fails. As a result, the failure rate is

$$\frac{\Delta N}{N \cdot t} \times 10^9 = \frac{10^9}{2500 \times 1000} = 400 FIT$$

The average of R(t):

Owing to its exponential distribution:

$$E(R(t)) = \int_0^\infty R(t)dt = \int_0^\infty exp(-\lambda t)dt = \frac{1}{\lambda} \tag{4-17}$$

There are different reliability concept of Repairable systems and Unrepairable Systems: when it is Repairable systems, MTBF is defined as the average of all possible time intervals for the system to deviate from its metrics when t tends to infinity.

$$MTBF = m = \int_0^{t \to \infty} tf(t)dt = \int_0^{t \to \infty} R(t)dt = \frac{1}{\lambda} \tag{4-18}$$

Unrepairable system: The average time before failure is MTTF. It is compared with the average time before the first failure MTTFF.

Average repair time:

$$MTTR = \frac{1}{N}\sum_{i=1}^N \Delta t_i \tag{4-19}$$

Availability:

$$A = \frac{MTBF}{MTBF+MTTF} \tag{4-20}$$

Some normal Lifetime distribution functions are in the following:

● Exponential failure distribution: electronic measurement system and components, normal working conditions.

● Normal failure distribution: some worn systems and components in area 3 of lifetime basin plot.

● Rayleigh distribution: When it comes to increased stress and accelerated wear.

● Gamma Distribution: Aging, wear, redundancy

● Weber distribution: At the same time there is aging, wear, over-load.

Example 4-9:

An effective life of a component with a failure rate of $0.0001h^{-1}$ is 1000 hours. If the failure distribution can be assumed to be exponential distribution, try to calculate MTBF and reliability in 10 hours,100 hours and 1000 hours.

Solution 4-9:

$$MTBF = \frac{1}{10^{-4}} = 10000(h)$$
$$R(10) = \exp(-\lambda t) = \exp(-10^{-4} \times 10) = 0.999$$
$$R(100) = \exp(-\lambda t) = \exp(-10^{-4} \times 100) = 0.990$$
$$R(1000) = \exp(-\lambda t) = \exp(-10^{-4} \times 1000) = 0.900$$

Example 4-10:

A circuit consists of 20 resistors with a failure rate of $10*10^{-9}$ for each component and 30 welds for each connection failure rate of $100*10^{-9}$. Calculate total inefficiency and MTBF if it can be assumed to be exponential distribution.

Solution 4-10:

$$\lambda_{total} = 20 \times 10 \times 10^{-9} + 30 \times 100 \times 10^{-9} = 3200 \times 10^{-9}(h)$$
$$MTBF = \frac{1}{\lambda_{total}} = \frac{1}{3200 \times 10^{-9}} = 35.67 years$$

The part is the smallest constituent unit component of the system. A small device may have several parts. Parts consist of a few components that can be used as part of a larger system to perform a function or task. The system is defined as a structure consisting of different types of equipment and it can complete the large special task in ISO9000 standards (Reliability can be defined as quality over time).

Availability:

$$A(t) = \frac{MTBF}{MTBF+MTTF} \tag{4-21}$$

The uptime maintenance rate is defined as the number of times that maintenance work is carried out per unit of time. Its derivative is MTTR.

For example, the maintenance rate of the system is $0.1h^{-1}$, then its MTTR

$$MTTR = \frac{1}{\mu} = 10(h)$$

If the MTTR of a computer system is 5 hours and the MTBF is 5,000 hours, the integrity rate

$$A(t) = \frac{MTBF}{MTBF + MTTF} = 0.999$$

The Boolean algebra of reliability:

C=A+B means: "If the system A or B or both systems are working, then system C work"

C=AB means "if both system A and B work, then the system C Works" these Boolean algebra Operation Law: it follows the algebra calculation principle, AA=A, A$\bar{A}$=0. And Morgan's laws:

$$\overline{A \cdot B} = \bar{A} + \bar{B} \tag{4-22}$$

$$\overline{A + B} = \bar{A} \cdot \bar{B} \tag{4-23}$$

Example 4-11: Compare the reliability between system a and system b in Fig. (4-5).

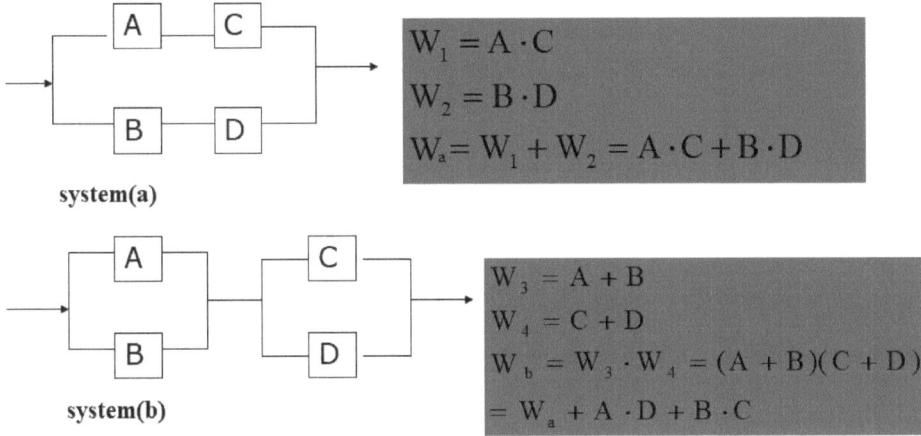

Fig. (4-5). The reliability of system (a) and system (b).

Solution 4-11:

The reliability of serial system is calculated by Equation (4-24).

$$R_{Serial} = \prod_{i=1}^{m} R_i \qquad (4\text{-}24)$$

If the distribution is exponential distributions:

$$R_{Serial} = \prod_{i=1}^{m} R_i(t) = \prod_{i=1}^{m} \exp(-\lambda_i(t) = \exp(-\lambda_s t) \qquad (4\text{-}25)$$

$$\lambda_s = \sum_{i=1}^{n} \lambda_i \qquad (4\text{-}26)$$

$$MTBF = 1/\sum_{i=1}^{n} \lambda_i \qquad (4\text{-}27)$$

The reliability of parallel system and its MTBF is calculated using the following Equations.

Its failure rate

$$F(A_{parallel}) = \prod_{i=1}^{m} (1 - P(A_i)) \qquad (4\text{-}28)$$

Its reliability rate

$$R(A_{parallel}) = 1 - \prod_{i=1}^{m} (1 - P(A_i)) \qquad (4\text{-}29)$$

If the distribution is exponential distributions:

$$R_{parallel} = 1 - \prod_{i=1}^{m} \exp(-\lambda_i(t) \qquad (4\text{-}30)$$

$$MTBF = \int_0^{\infty}{}^{\infty}(1 - \prod_{i=1}^{m} \exp(-\lambda_i(t)) \, dt \qquad (4\text{-}31)$$

4.3.2. Reliability Program Plan and Assessment

The objectives of reliability engineering are:

● To prevent or to reduce the occurrences of failures.

● To find the reason for failures.

● To cope with failures.

● To analyze and process reliability data.

The diagram of reliability plan is shown in Fig. (**4-6**).

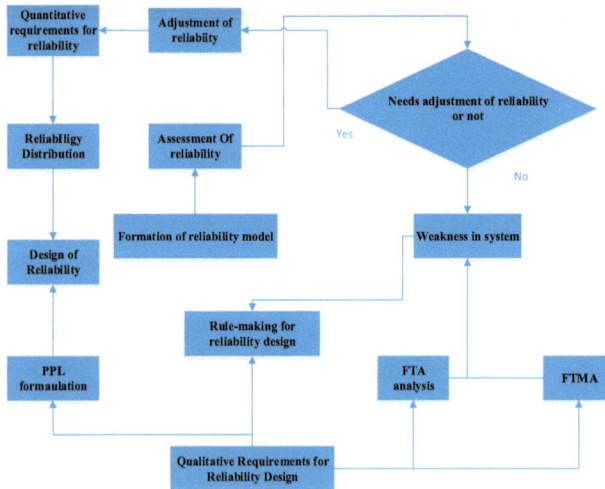

Fig. (4-6). The reliability program plan diagram [9].

1) Reliability allocation

There are two kinds of distribution method:

★ The equal distribution method: every subsystem has an equal failure rate.

$$R = \prod_{i=1}^{n} R_i \qquad \text{(4-32)}$$

★ The distribution method of aeronautical Radio company

1. The goal is to meet the following formula:

$$\sum_{i=1}^{n} \lambda_i \le \lambda \qquad \text{(4-33)}$$

2. Calculates the weighted factor W_r based on the prior knowledge to estimate the inefficiency of each sub-system λ_i

3. The weighted factor is calculated from the following formula:

$$W_r = \frac{\lambda_r}{\sum_{i=1}^{n} \lambda_i} \qquad \text{(4-34)}$$

4. Distributes inefficiency to each sub-system

Example 4-12:

A system consists of 3 subsystems, and the failure rates of known 3 subsystems are $\lambda_1 = 0.003$, $\lambda_2 = 0.001$, $\lambda_3 = 0.004$ respectively. The 20-hour reliability of the system is set at 0.9, which is distributed according to the principle of the reliability calculation of the Aviation Radio Company.

Solution 4-12:

(1) The failure of the system

$$R(20) = \exp(-\lambda \cdot 20) = 0.9 \Longrightarrow \lambda = 0.005$$

(2) The weighted factor W_r

$$W_1 = \frac{\lambda_1}{\lambda_1 + \lambda_2 + \lambda_3} = 0.375; \quad W_2 = \frac{\lambda_2}{\lambda_1 + \lambda_2 + \lambda_3} = 0.125;$$

$$W_3 = \frac{\lambda_3}{\lambda_1 + \lambda_2 + \lambda_3} = 0.5;$$

(3) The failure rate of subsystem

$$\lambda_1' = W_1\lambda = 0.375 \times 0.005 = 0.001875$$
$$\lambda_2' = W_2\lambda = 0.125 \times 0.005 = 0.000625$$
$$\lambda_3' = W_3\lambda = 0.5 \times 0.005 = 0.0025$$

(4) The reliability of subsystem

$$R_1(20) = 0.96$$
$$R_2(20) = 0.99$$
$$R_3(20) = 0.95$$

In Fig. (**4-12**), the PPL is Component Optimization Checklist.

2) Reliability prediction method

It is expected to predict reliability based on the reliability of each component of the product in predicting the reliability of the product under specified operating conditions. It is expected that the reliability of the system will be inferred according to the reliability of the components, components, and sub-systems of the system

The normal method includes:

A. Similar product method: The reliability of mature products is estimated by using the reliability of products like those of the product, and the reliability data of the proven product mainly comes from the field usage statistics and laboratory test results. The expected accuracy depends on the similarity of the product and the accuracy of the reliability data of the mature product.

B. Rating Forecast method: In the absence of product reliability data, you can ask experts who are familiar with the product, have practical experience in engineering, according to several major factors affecting product reliability (such as complexity, technical maturity, importance, and environmental conditions) to score (the score of each factor between 1~10, the higher the difficulty score higher), The reliability index is then assigned to each sub-system or component according to the result of the score.

Example 4-13:

The system (its MTBF is 500h), it is composed of A/B/C/D four parts, the score of every part is in Table **4-2**, please locate the failure rate and MTBF.

Table 4-2. The score of every part and their calculations.

parts	complex	Technology maturity	important	environment	score	Score parameter	Failure rate	MTBF
A	8	9	6	8	3456			
B	5	7	6	8	1680			
C	5	6	6	5	900			
D	6	6	8	5	1440			
SUM								

Solution 4-13:

parts	complex	Technology maturity	important	environment	score	Score parameter	Failure rate	MTBF
A	8	9	6	8	3456	0.4623	0.00092	1082
B	5	7	6	8	1680	0.2247	0.00045	2225
C	5	6	6	5	900	0.1204	0.00024	4153
D	6	6	8	5	1440	0.1926	0.00038	2596
SUM					7476	1	0.002	500

4.3.3. Reliability Analysis

1) Fault tree analysis of Reliability Analysis:

The fault tree definition a fault tree is a logical diagram that indicates which components of the product are faulty or external events or their combination will cause the product to occur with a given failure. A fault tree is a logical causality diagram in which the elements of an event and a logical gate event are used to describe a system and meta, part failure, and the state logic gate connects events to represent the logical relationship between events [11].

Fault Tree Analysis (FTA) through the analysis of hardware, software, environment, human factors that may cause product failure, draw fault tree, to determine the cause of product failure of various possible combinations and/or probability of its occurrence.

Example 4-14: The simple FTA analysis of Titanic shipwreck

Solution 4-14:

It is shown in Fig. (**4-7**).

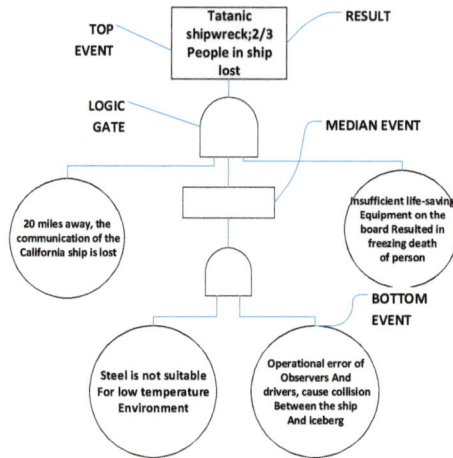

Fig. (4-7). The simple FTA solution of Tatanic shipwreck.

Another example is FTA of motor, it is shown in Fig. (**4-8**).

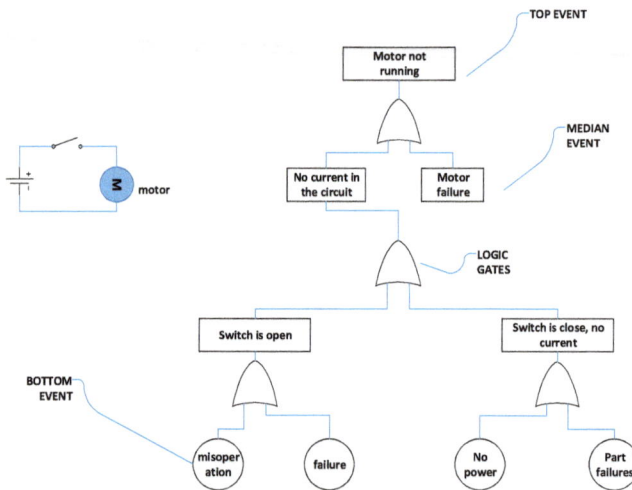

Fig. (4-8). The FTA analysis of "motor not running.

2) FEMA

Fault mode Impact analysis (Failure Mode and Effects Analysis, shorthand for FMEA) is an analysis of all possible failure patterns and all possible impacts on the system for each product in the system, and according to the severity of each failure mode, An inductive analysis method for classification of detection difficulty and frequency of occurrence [12].

4.3.4. Reliability Design

1) EMC design

EMC Design includes electrostatic, lightning, surge, fast transient pulse, power ripple, and instantaneous drop, conduction interference, radio frequency electromagnetic field protection design.

The EMC hardware method includes filtering, grounding, shielding and isolation. The EMC software method includes data filtering, software watchdog, instruction redundancy and program trap.

Example 4-15: The EMC/EMI model from Integrated Circuits [13]

Solution 4-15:

To copy with the EMC/EMI issues in IC, there are many methods. For example, use methods of a full-chip simulation.

The other method had been proposed, for example, the segmentation technique using a proper equivalent circuit model.

The simple and effective tool is important in this issue. The EMC performance of a complex system made by an IC, package, and PCB is often using the IBIS model, based on the "Input/output buffer information specifications".

But it lacks the inclusion of main phenomena, such as the switching noise and the load effect of the PCB.

So, the International Electrotechnical Commission proposed the method of the Integrated Circuit Emission Model (ICEM)s.

The conceptual scheme of ICEM is shown in Fig. (**4-9**), it includes:

• internal activity(IA) describes the core and the I/O buffers;

• the Package;

• the printed circuit board (PCB).

• a generic load [13].

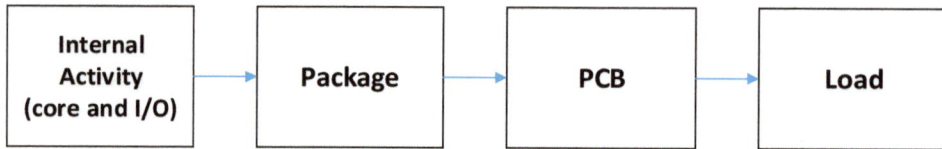

Fig. (4-9). The conceptual scheme of the proposed circuit model.

Example 4-16: Processing Principles of Various Groundlines

Solution 4-16: Processing Principles of Various Groundlines includes:

● Principle of One Point Grounding for Low Frequency Circuits;

● Multipoint Grounding Principle for High Frequency Circuits;

● Separation of strong ground wire and signal ground wire;

● Distribution of analog signal ground wire and digital signal.

2) Tolerance Design

Tolerance design studies the fluctuation range of performance indicators. It includes components, components, circuits, software tolerance. Circuit Tolerance Analysis is the Prediction Technology of Circuit Performance Parameter Stability [14].

The reason for drift degradation of the device and circuit are: Ignoring tolerance, Environment conditioning effect, Degradation effect. So, the design of intelligent instrument includes System Design (Functional Design), Parameter Design (Quality Design), and Tolerance Design (Pre-design, Sensitivity Analysis).

Tolerance design is to determine the appropriate tolerance of each parameter.

The basic idea is that according to the contribution of each parameter to the product performance, it is necessary to give a smaller tolerance to the parameters that have a greater impact from the economic point of view.

There are many ways to realize tolerance design, such as Worst-case, Root-Su--Squares and Simulation.

Example 4-17: An example of tolerance design (prediction of Tolerance Range)

Install four parts in an assembly ring. As shown in the figure below, the target value of Gap is required to be T=0.016 and the fluctuation range is as small as possible. It is known that the current 1-4 parts obey the technical specifications 1.225 (+0.003), and the assembly ring obeys the technical specifications 4.916 (+0.003). Question: Does the target value of the system meet the requirements? What is the tolerance range?

Solution 4-17:

A. Worst-case method: According to the analysis thought of extreme value analysis method:

Nominal value of assembly ring = 4.916 tolerance = 0.003

Nominal Value of Part 1=-1.225 Tolerance=+0.003

Nominal Value of Part 2=-1.225 Tolerance=+0.003

Nominal Value of Part 3=-1.225 Tolerance=+0.003

Nominal Value of Part 4=-1.225 Tolerance=+0.003

From this we can get:

The nominal value of clearance = 0.016 total tolerance = 0.015

Minimum clearance = 0.001

Maximum clearance = 0.031

That is to say, the target value of the system meets the requirement, and the tolerance range of the system is [0.001, 0.031].

B. Root-Sum-Squares method: Statistical square tolerance method is based on the hypothesis that most parts show normal probability distribution within their tolerance range. When the system composed of them is linearly related to each part, the distribution of the system can also be expressed by a normal distribution or approximate normal distribution.

Under the condition of the same mechanical system, according to the definition formula of statistical square tolerance method, the total tolerance of clearance=

Minimum clearance = 0.016-0.0067 = 0.0093

Maximum clearance = 0.016 + 0.0067 = 0.0227

That is to say, the tolerance range of the system becomes [0.0093, 0.0227].

3) Potential Circuit Analysis (SCA)

The so-called latent circuit refers to the unwanted path generated in the circuit under certain conditions. Its existence can cause an abnormal function or inhibit normal function. It includes the following cases:

Sneak path: Unexpected Current Path or Unexpected Logical Sequence Flow;

Potential timing: An unexpected event or sequence of events conflicts.

Latent instructions: The system operation indication is not clear or the wrong display will cause the system or the operator to make the unexpected action.

Latent signs: Faulty or imprecise system functionality flags (such as system input, control, display bus, *etc.*) can cause operators to operate the system incorrectly.

It often uses the topology tree methods, the topology tree includes 5 structure, shown in Fig. (**4-10**) [15].

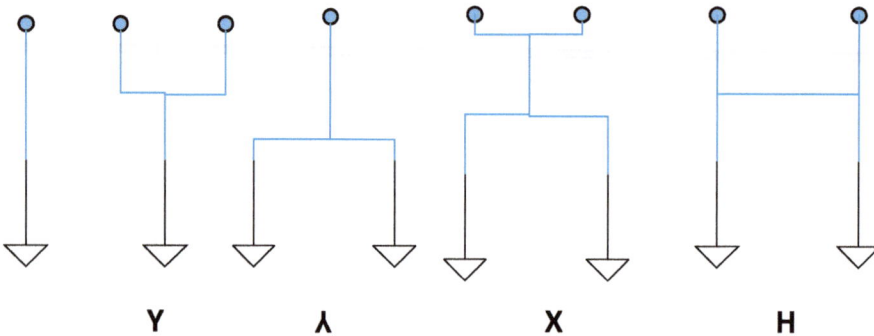

Fig. (**4-10**). The five-base topology of SCA topology tree.

Example 4-18: SCA1: the signal line passes through the same group of ground lines.

Solution 4-18:

In the 51st launch of American Ruby Rocket, the tail plug (TB) was disconnected 29 ms earlier than the drop plug (TC) to make the point. The fire signal passes through the light and then through the sub-channel (emergency shutdown), Indicate relay coil and restraint diode, TC, shutdown coil to ground, resulting in the rocket just take off and shutdown, back to the launch pad. It is shown in Fig. (**4-11**).

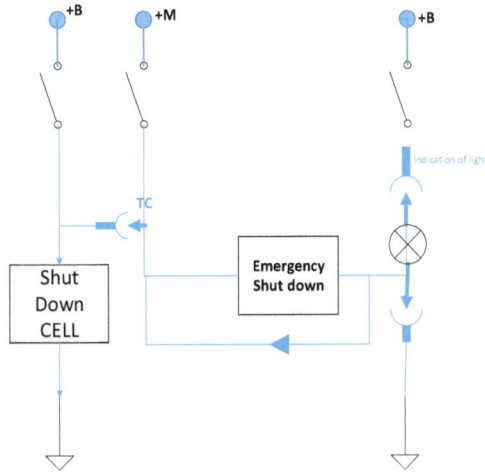

Fig. (4-11). One example of SCA.

Parallel shunt circuit operates when the branches are unequal, it may lead to an overload of each branch circuit, thus reducing the reliability life. Parallel shunt circuit operates when the branches are unequal. It is shown in Fig. (**4-12**) [15].

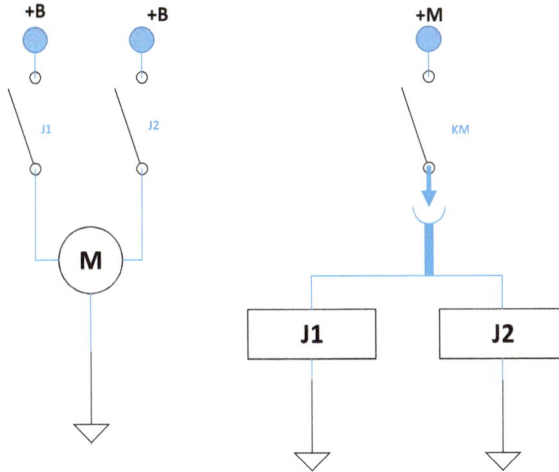

Fig. (4-12). Parallel shunt example of SCA.

On-site testing: At present, the DC motor cannot cut off electricity because of the sintering of contacts. The reason for sintering is that the actual on-off time of Er and J2 is different. For inductive loads like motors, the contacts that are switched on early must withstand the full Pu start current. It is required to select a circuit breaker according to 0.3 reduction coefficient. Electric machinery Rated current is 6A.J1, J2 contacts rated current are all 12A. The load-bearing parameters of a single branch selected according to 0.3 of the follow up shall be as follows:6A/0.3=20A.

4) Simplified design

System reliability is a function of its complexity. Simplified design refers to the simplification of the number and complexity, accuracy, dimensional accuracy, shape and position requirements, structure or whole component/system requirements of components in the product design process, to achieve the simplest state on the premise of guaranteeing the performance requirements.

5) Derating design

The reliability reduction design of an intelligent instrument mainly refers to that the stress (electrical stress and temperature stress) of components constituting the instrument is lower than the rated value of its design to delay the degradation of parameters and prolong the reliability of its use.

6) Redundancy and tolerance design

The concept of redundancy, strictly speaking, is to use multiplied components to participate in the control, to minimize the loss caused by the accident of the control equipment.

A. Processor Redundancy: When the main processor (called heat engine) fails, the standby processor (called standby machine) automatically goes into operation and takes over control. Due to the switching mechanism and speed, it can be divided into Cold Redundancy (Cold Standby) and Thermal Redundancy (Hot Standby).

B. Communication redundancy: The most common is the dual-channel communication cable. For example, two-cable Profibus communication or two-cable ControlNet communication. Simple communication redundancy, single module, and double cable mode, two sets of single module and single cable duplex mode.

C. I/O Redundancy: Redundancy of I/O is the most difficult to implement compared to processor and communication. Usually I/O redundancy is less used and the cost increases more.

Soft Redundancy: Generally, it refers to the cold standby of processors. Cold standby is switched by software. Processors are usually used in pairs, one for use and one for standby. When the main processor fails, switch to the standby processor by software. Slow speed and low cost.

Hard Redundancy: Generally, it refers to the hot standby of processors. Hardware switching for a hot standby system.

Voting system: The security control system is often used. 1:1:1 mode. Special functions can be realized, such as 1 with 2 spares, or 3 with 2 votes. Typical products are GE's S90-70. Turbine protection systems in thermal power plants and ESD systems in the petrochemical industry are all in this way.

7) Signal Integrity Design

When the signal can't work normally or the signal quality can't make the system work stably for a long time, the problem of signal integrity arises. When the operating frequency of the circuit is above 100M Hz, this problem needs to be considered.

• Signal transmission quality in a single network

• Crosstalk between multiple networks

• Track Collapse in Power Supply and Ground Distribution

• EMI and radiation from the whole system

PROBLEM

4-1. Briefly describe the relationship and difference between modern design concept and traditional design concept.

4-2. The basic structure of data acquisition system is described.

4-3. What are the basic functions of data acquisition system?

4-4. The sampling period is related to those factors. How to choose the sampling period?

4-5. What factors should be considered when choosing a computer for a data acquisition system

4-6. How many kinds of keyboards are made up? What are their characteristics? What are the ways to de-jitter keys?

4-8. What are the main problems of keyboard interface?

4-9. What are the two commonly used driving modes for LCD? What is the principle?

4-10. Discuss the working principle of common touch screen

4-11. How many transmission modes are there in USB? Brief description of the characteristics of various transmission modes

4-12. What are the main factors to be considered in the design of possibilities?

4-13. Brief description of the process of fault tree analysis and the main contents of each step

4-14. Brief description of FEMA effect standard and method

4-15. Four thousand instruments are required to have MTBF = 1000 hours. If it is run for 1000 hours, how many instruments are allowed to fail at most?

4-16. If three instruments with reliability of 0.9 are connected in series to form a system, what is the reliability of the system?

4-17. If it is known that a system consists of four subsystems, the failure rates of the subsystems are 0.003, 0.0025, 0.004 and 0.001. The 100-hour reliability of the system is 0.7. The reliability of four subsystems is assigned according to the method of equalization and the method of Aeronautical Radio company. The 100-hour reliability of each subsystem is calculated separately

REFERENCES

[1]   "Accuracy" available from https, https://en.wikipedia.org/wiki/

[2]   B.G. Wang, and Z.B. Pu, *Design of measurement and control instrument.* 2nd ed. The China Machine Press: Beijing, 2009.

[3]   A. Sachenko, *Error compensation in an intelligent sensing instrumentation system.* IEEE Instrumentation & Measurement Technology Conference IEEE, 2001. [http://dx.doi.org/10.1109/IMTC.2001.928201]

[4]   "Data quality" available from, https://en.wikipedia.org/wiki/

[5]   "software test" available from, https://en.wikipedia.org/wiki/

[6]   D.W. Hoffmann, *Software-Test.* Software-Qualität, 2013. [http://dx.doi.org/10.1007/978-3-642-35700-8_4]

[7]   D.F. Chen, and Q. Lin, *The Intelligent Instrument.* The China Machine Press: Beijing, China, 2014.

[8]   "Reliability" available from, https://en.wikipedia.org/wiki/

[9]   L. Fu, C.J. Deng, and L,. Nie, *Theory, Design and Application of Intelligent Instruments* Southwest Jiao tong University Press: China, Chengdu, 2014.

[10] D. Dubois, and H. Prade, *Possibility Theory: Qualitative and Quantitative Aspects.* Quantified Representation of Uncertainty and Imprecision, 1998.

[11] "FTA" available from https, https://en.wikipedia.org/wiki/

[12]   "FEMA" available from, https://en.wikipedia.org/wiki/

[13]   Accurate Models for Evaluating the Direct Conducted and Radiated Emissions from Integrated Circuit, *Appl. Sci. (Basel),* vol. 8, p. 477, 2018.
[http://dx.doi.org/10.3390/app8040477]

[14]   "Tolerance design" available from, https://en.wikipedia.org/wiki/

[15]   D. Yan, *Sneak Passage Analysis Technology.* Missiles & Space Vehicles, 2000.

CHAPTER 5

The Data Communication and Input-output Technology of Intelligent Instrument

Abstract: In this chapter, the IIC, SPI, EPA communication technology will be discussed, and the input-output technology includes HMI, OLED will be introduced and analyzed.

Keywords: Data communication, Display technology, EPA, HMI, IIC, OLED, SPI.

5.1. THE DATA COMMUNICATION TECHNOLOGY

5.1.1. IIC Bus

IIC (Inter-Integrated Circuit) literally means between integrated circuits. It is a serial communication bus. It uses multi-master-slave architecture. Philips in the 1980s used it to connect motherboards, embedded systems or mobile phones.

I2C serial bus usually has two signal lines, one is bidirectional data line SDA, the other is clock line SCL. All serial data SDA connected to I2C bus devices are connected to SDA of the bus, and the clock line SCL of each device is connected to SCL of the bus.

To avoid the confusion of bus signals, it is required that the output of each device connected to the bus must be drain open circuit (OD) output or collector open circuit (OC) output. The serial data line SDA interface circuit on the device should be bidirectional, the output circuit should be used to send data to the bus, and the input circuit should be used to receive data on the bus. The serial clock line should also be bidirectional. As the host of controlling bus data transmission, on the one hand, the clock signal should be sent through the SCL output circuit, on the other hand, the SCL level on the bus should be detected to decide when to send the next clock pulse level; as the slave receiving the host's command, the SCL on the bus should be used as the slave receiving the host's command. Signal sends or receives signals on SDA, and can also send low-level signals to SCL lines to extend bus clock signal cycle. When the bus is idle, the pull-up resistance Rp keeps the SDA and SCL lines at high levels because all the devices are open-leak outputs. The low output level of any device will make the corresponding bus signal line lower, that is to say, SDA of each device is "with" relationship, SCL is also "with" relationship.

Bus has no special requirements for the manufacturing process and level of equipment interface circuit (NMOS, CMOS can be compatible). The data transmission rate on I2C bus can be as high as 100,000 bits per second and over 400,000 bits per second in high-speed mode. Besides, the number of devices allowed to connect on the bus is limited to a capacitance of less than 400 pF.

The operation of the bus (data transmission) is controlled by the host. The so-called host refers to the device that transmits the start data (sending out the start signal), sends out the clock signal and sends out the stop signal at the end of the transmission. Usually the host is a microprocessor. The device being visited by the host is called a slave device. To communicate, each device connected to the I2C bus has a unique address to facilitate host access. Data transmission between host and slave can be transmitted from host to slave or from slave to host. Every device that sends data to the bus is called a transmitter, and the device that receives data from the bus is called a receiver.

I2C bus allows the connection of multiple microprocessors and various peripheral devices, such as memory, LED and LCD driver, A/D and D/A converter, *etc*. To ensure reliable data transmission, the bus can only be controlled by one host at any time. Each microprocessor should send start-up data when the bus is idle.

To properly resolve the conflict between multiple microprocessors sending start-up data at the same time (bus control) and decide which microprocessor controls the bus, there are some solutions. For example, that is the problem: The I2C bus allows devices with different transmission rates to be connected, to solve this problem, the synchronization method is used. Synchronization of clock signals between multiple devices is called synchronization.

Data on SDA line must be stable during the period of "high" clock. Only when the clock signal on SCL line is low, the "high" or "low" state on data line can be changed. Each byte exported to the SDA line must be 8 bits, and each byte transmitted is unrestricted, but each byte must have a response ACK. If a receiver cannot receive a complete byte of another data before completing other functions, such as an internal interrupt, it can keep the clock line SCL low to cause the sender to enter a waiting state; when the receiver is ready to receive other bytes of data and release the clock SCL, data transmission continues Fig. (**5-1**).

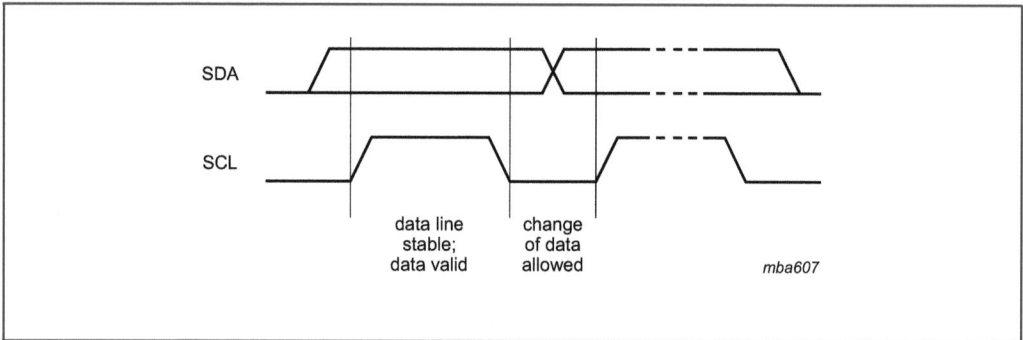

Fig. (5-1). Bit transfer on the I2C bus.

In the process of I2C bus transmission, two specific conditions are defined as start and stop conditions Fig. (**5-3**). When SCL remains "high", SDA changes from "high" to "low" as the starting condition; when SCL remains "high" and SDA changes from "low" to "high", SDA becomes the stopping condition. Both start and stop conditions are generated by the main controller. Start and stop conditions can be easily detected using hardware interfaces. Microcomputers without such interfaces must sample SDA at least twice per clock cycle to detect such changes Fig. (**5-2**).

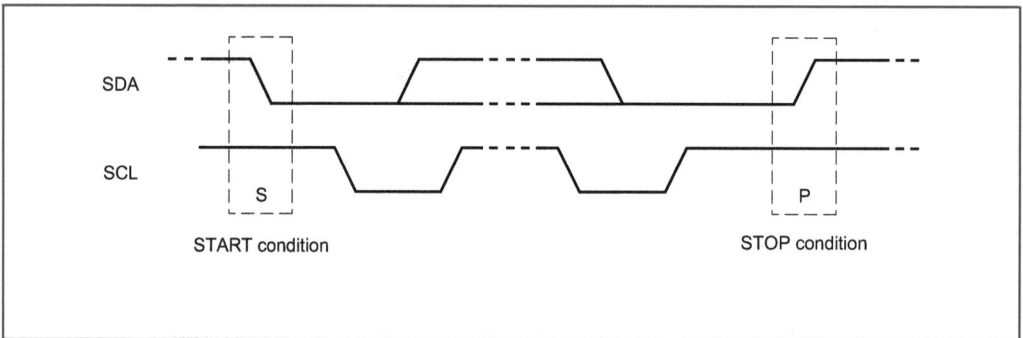

Fig. (5-2). The start and stop conditions.

Every byte put on the SDA line must be eight bits long. The sequence of Data transferred is the Most Significant Bit (MSB) first Fig. (**5-3**).

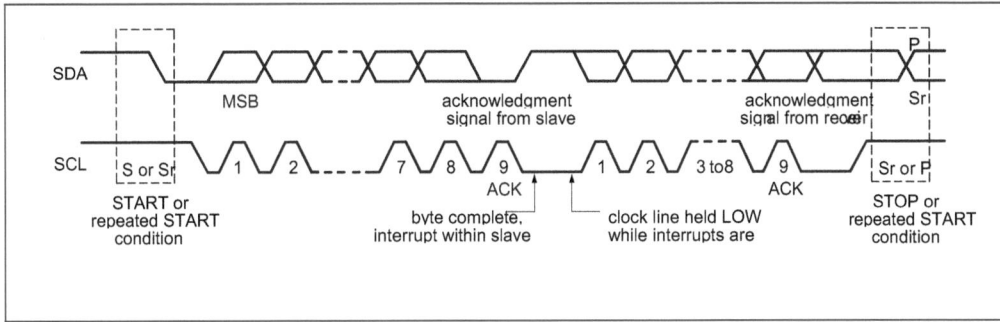

Fig. (5-3). The data communication of I2C bus.

Data transmission must follow a response. The clock pulse corresponding to the reply is generated by the main controller, and the sender must pull down the SDA line during the reply. When the addressable controlled device fails to respond, the data remains high and the master produces a stop condition to terminate transmission. In the process of transmission, when the master receiver is used, the master receiver must send a data end signal to the controlled transmitter, so that the controlled transmitter releases the data line to allow the master to produce a stop condition.

The format of data is shown in Fig. (5-4).

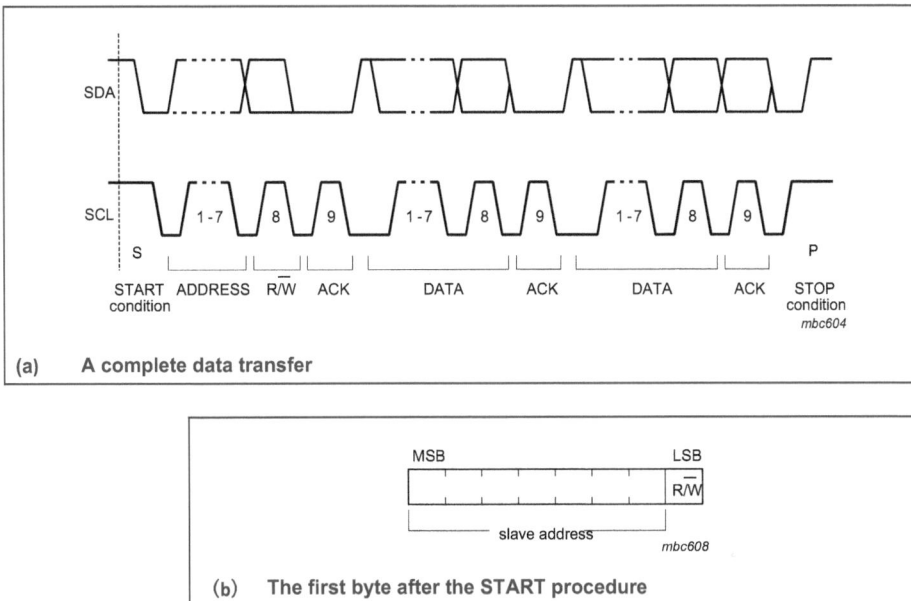

(a)   **A complete data transfer**

(b)   **The first byte after the START procedure**

Fig. (5-4). The first byte and complete data transfer of I2C bus.

First is the START condition (S);

Second, send a slave address. IIC address has seven bits long, the lowest bit is data direction bit (R/W) ;

— a 'zero' indicates a transmission (WRITE),

a 'one' indicates a request for data (READ).

Each byte must be guaranteed to be 8 bits long. In data transmission, the highest bit (MSB) is transmitted first, and each byte transmitted must be followed by a reply bit (that is, a frame has 9 bits). If the slave response signal is not received for a period of time, the slave automatically considers that the data has been received correctly.

IIC Writing: When the MCU writes, it sends the device's 7-bit address code firstly and write direction bit "0" (a total of 8 bits, that is, a byte), releases the SDA line after sending, and generates the 9th clock signal on the SCL line. After confirming that the selected memory device is its address, a response signal is generated on SDA line as a response, and the MCU can transmit data after receiving the response. When transmitting data, the MCU first sends a byte to the first address of the memory. After receiving the response from the memory device, the MCU sends data bytes one by one, but after each byte, it waits for the response. The in-chip address of AT24C series automatically adds 1 after receiving each data byte address. Within the limit of "the number of bytes loaded once" of the chip, only the first address needs to be input. When the number of bytes loaded exceeds the "number of bytes loaded at one time" of the chip, the data address will be "rolled up" and the previous data will be overwritten.

IIC read: The MCU transmits the device's 7-bit address code and writes direction bit "0" ("pseudo-writing"), releases the SDA line after sending, and generates the 9th clock signal on the SCL line. After confirming that the selected memory device is its own address, a response signal is generated on the SDA line as a response.

Then, a byte is sent to read out the first address of the memory area of the device. After receiving the response, the MCU will repeat the starting signal and send out the device address and read direction bit ("1"). After receiving the device response, the data byte can be readout. For each byte readout, the MCU will reply to the response signal. When the last byte of data is read out, the MCU should return to the "non-response" (high level) and send a stop signal to end the reading operation. This process is shown in Fig. (5-5).

IIC WRITE

SDA LINE

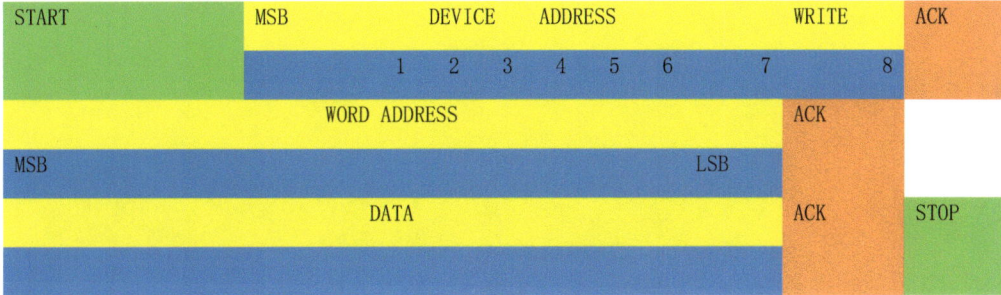

IIC READ

SDA LINE

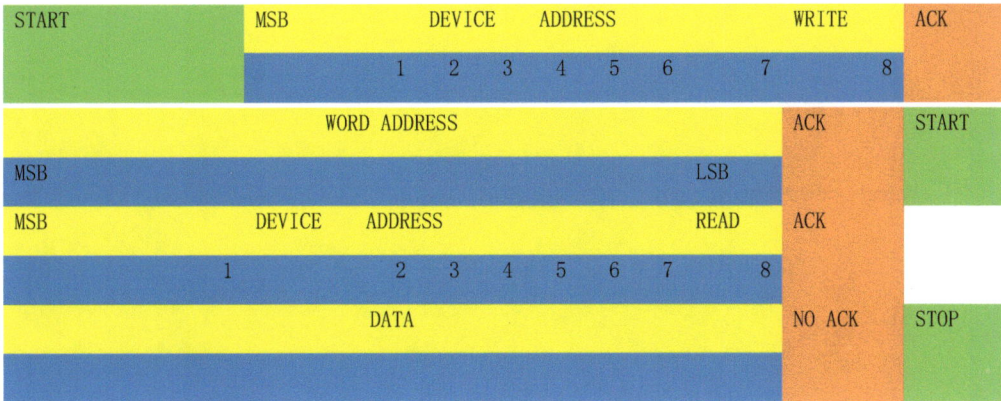

Fig. (5-5). The read and write procedure of I2C bus.

Example 5-1: C code of IIC

Solution 5-1:

void init() //initiation

{

SCLK=1;

delay();

```
SDATA=1;

delay();

}

void IICstart() //start signal

{

SDATA =1;

delay();

SCLK =1;

delay();

SDATA =0;

delay();

}

void IICstop() //stop signal

{

SDATA =0;

delay();

SCLK =1;

delay();

SDATA =1;

delay();

}

uchar IICreadbyte() //read one byte

{

uchar i,j,k;
```

```
SCLK =0;
delay();
SDATA =1;
for(i=0;i<8;i++)
{
SCLK =1;
delay();
if(SDATA ==1)
j=1;
else
j=0;
k=(k<<1)|j;
SCLK =0;
delay();
}
delay();
return k;
}
uchar IICwritebyte(uchar data) //write one byte
{
uchar i,temp;
temp=date;
for(i=0;i<8;i++)
{
```

```
temp=temp<<1;

SCLK =0;

delay();

SDATA=CY;

delay();

SCLK =1;

delay();

}

SCLK =0;

delay();

SDATA =1;

delay();

}

void IICwrite_add(uchar address,uchar info) //write data to an address

{

IICstart ();

writebyte(0xa0);

response();

writebyte(address);

response();

writebyte(info);

response();

IICstop ();

}
```

```
uchar IICread_add(uchar address) //read one byte from an address

{

uchar dd;

IICstart ();

writebyte(0xa0);

response();

writebyte(address);

response ();

IICstart ();

writebyte(0xa1);

response ();

dd=readbyte();

IICstop ();

return dd;

}

void response() //acknowledge signal

{

uchar i=0; SCLK =1;delay();

while((SDATA ==1)&&(i<255))

i++;

SCLK =0;

delay();

}
```

5.1.2. SPI Bus

SPI interface is Motorola's first full duplex three-wire synchronous serial peripheral interface, which adopts Master Slave architecture and supports multi-slave mode applications, generally only single Master.

The clock is controlled by Master. Under the clock shift pulse, the data is transmitted bit by bit, with the high bit ahead and the low bit behind (MSB first). The SPI interface has two one-direction data lines, which are full-duplex communication. The data rate in current applications can reach several Mbps.

There are four signal lines in SPI interface: device selection line, clock line, serial output data line and serial input data line. The structure of SPI is shown in Fig. (5-7) [2].

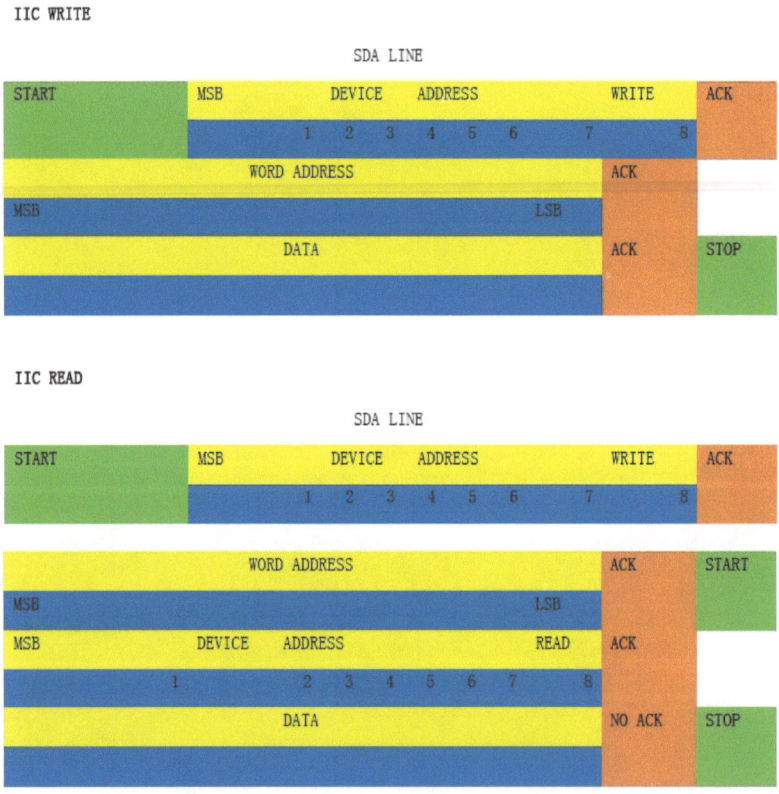

Fig. (5-6). The structure of SPI bus.

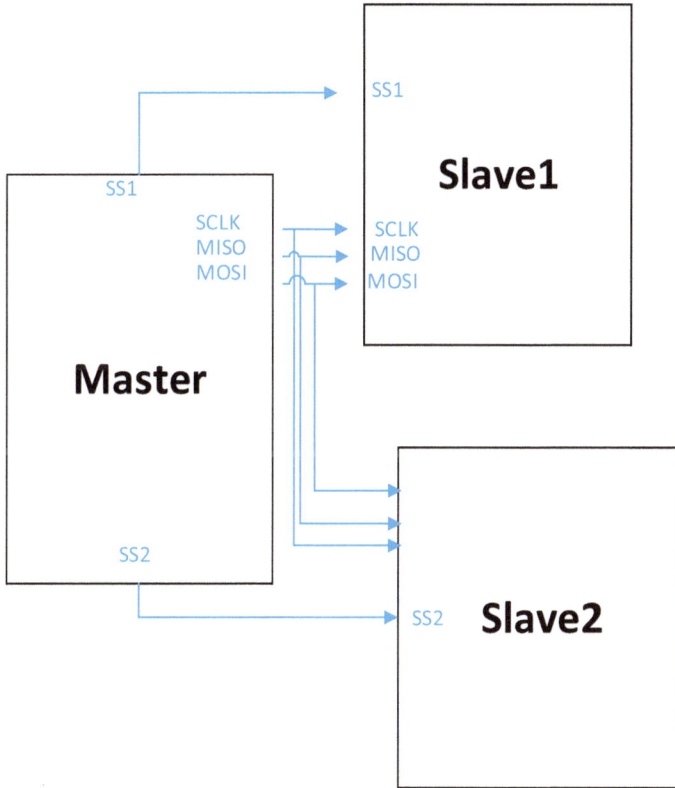

Fig. (5-7). The structure of SPI bus.

(1) MOSI: Main device data output, data input from device.

(2) MISO: Data input of the main device and data output of the slave device.

(3) SCLK: clock signal, generated by the main device.

(4) /SS: Slave device enabling signal controlled by the main device.

The SPI interface is two simple shift registers in the internal hardware. The data transmitted is 8 bits. Under the enabling signal and shift pulse of the slave device generated by the main device, the SPI interface transmits bit by bit, with the high position in front and the low position in behind. As shown in the figure below, the data changes on the descending edge of SCLK and the ascending edge is stored in the shift register [3].

First, the host sends commands, then the slave prepares the data according to the host's commands. The host reads the data back in the next 8-bit clock cycle.

Example 5-2: Example code of touch screen IC ADS7843 [4]

Solution 5-2:

The ADS7843 is a 12-bit sampling Analog-to-Digital Converter (ADC) of resistance switches in a touch screen.

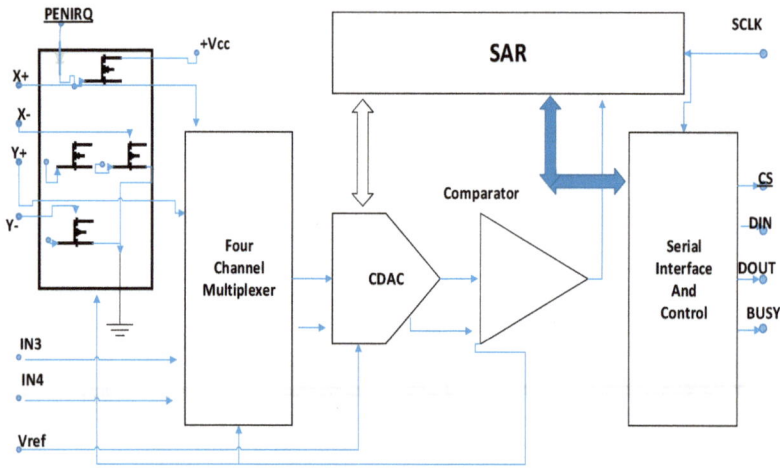

Fig. (5-8). The structure diagram of ads7843.

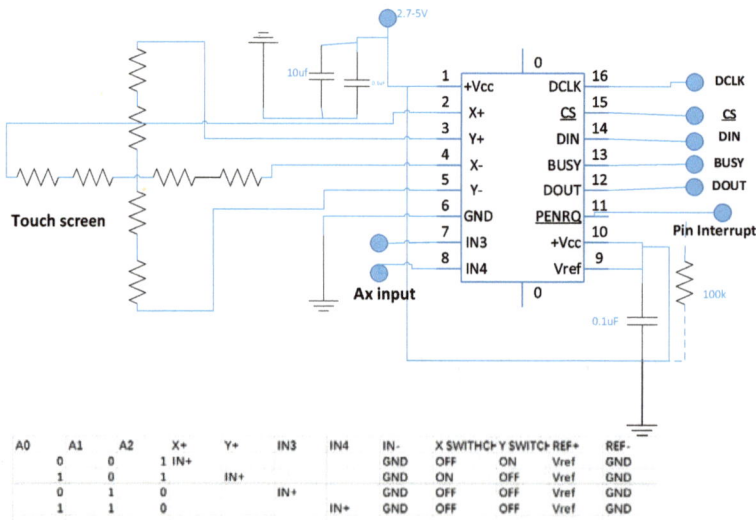

Fig. (5-9). The basic operation of ads7843.

A0	A1	A2	X+	Y+	IN3	IN4	IN-	X SWITCH	Y SWITCH	REF+	REF-
0	0	1	IN+				GND	OFF	ON	Vref	GND
1	0	1		IN+			GND	ON	OFF	Vref	GND
0	1	0			IN+		GND	OFF	OFF	Vref	GND
1	1	0				IN+	GND	OFF	OFF	Vref	GND

The first bit, the 'S' bit, must always be HIGH and indicates the start of the control byte.

The ADS7843 will ignore inputs on the DIN pin until the start bit is detected.

The next three bits (A2-A0) select the active input channel or channels of the input multiplexer. The Model bit determines the number of bits for each conversion, either 12 bits (LOW) or 8 bits (HIGH).

The C code to drive ADS7843 is in the following:

```c
//==========================================================================

// ADS7843

#include <reg51.h> //STC

#define unint unsigned int

#define unchar unsigned char

sbit DataCLK = P2^0; //

sbit CSelet = P2^1; //

sbit DataIN = P2^2; //

sbit BUSY = P2^3; //

sbit DataOUT = P2^4; //

sbit PENIRQ = P3^2; //

void Tranfer(char Data);

uint average(uint a [8]);

void ADS7843_start(void)

{

DataCLK =0;

CSelet =1;

DataIN =1;
```

```
DataCLK =1;

CSelet =0;

}

void ADS7843_wr(uchar datt)

{

uchar count;

DataCLK =0;

for(count=0;count<8;count++)

{

datt<<=1;

DataIN =CY;

DataCLK =0;

_nop_(); _nop_(); _nop_();

DataCLK =1;

_nop_(); _nop_(); _nop_();

}

}

uint ADS7843_rd(void)

{

uchar cnt=0;

uint datt=0;

for(cnt=0;cnt<12;cnt++)

{

datt<<=1;
```

DataCLK =1; _nop_();_nop_();_nop_();

DataCLK =0; _nop_();_nop_();_nop_();

if(DataOUT)

datt++;

}

return(datt);

}

5.1.3. EPA Bus

Overall framework features of EPA bus are:(1) Distributed architecture;(2) EPA Architectural Reference Model Complying with IEC 61499. This automation system consists of physical model and logical equipment. Its reference model includes: system model, equipment model, resource model. (3) Support vertical (information) integration and horizontal (automation) integration; (4) Providing overall solutions, Using micro-network segment structure(time-sharing transmission by configuration) [5].

EPA control system consists of an engineer station, operator station, EPA agent station, bridge, field equipment and system network (information management network, process control network, EPA field control layer network).

(1) System Model

The application process of the control system is distributed on one device or a group of devices. Through network interconnection, the distributed application operates jointly to realize the control of process objects.

(2) Equipment Model

It consists of devices that one or more resources on which local parts of distributed applications can be executed by combining these resources. These devices are called EPA physical devices in EPA standard, or EPA devices for short.

(3) Resource Model

It consists of interconnected software objects, which can communicate directly with each other within the device or can interconnect through EPA network. In IEC 61499, these software objects are called function blocks; in EPA standards, these software objects are called EPA blocks.

EPA blocks are atom-level elements in EPA systems, they are indivisible basic elements. They exist in an entity containing executable algorithms and have interconnectable external interfaces for input and output data. EPA block is the distributed function unit of EPA application.

IEC 61499 assumes that a function block (corresponding to EPA block) application process resides in a device consisting of different logic devices, or in a device consisting of different logic devices. These logic devices are formed by causal links between event flow and data flow. Events from the output side are combined with data from the associated EPA block. Input for another EPA block. For an EPA block application process consisting of one or more EPA logic devices in an EPA device, the communication relationship between its internal object modules can be defined by the EPA device manufacturer.

IEC 61499 assumes that a function block (corresponding to EPA block) application process resides in a device consisting of different logic devices, or in a device consisting of different logic devices. These logic devices are formed by causal links between event flow and data flow. Events from the output side are combined with data from the associated EPA block. Input for another EPA block. For an EPA block application process consisting of one or more EPA logic devices in an EPA device, the communication relationship between its internal object modules can be defined by the EPA device manufacturer.

(5) Wireless Access

EPA standard defines two types of WLAN devices, namely, WLAN EPA field devices and EPA access devices. Wireless LAN EPA access devices usually consist of a wireless LAN interface and an Ethernet interface. The EPA access devices load the IEEE 802.1d bridge protocol. It supports the following access: data exchange between EPA field devices directly, and wireless LAN EPA field devices connect to Ethernet through access devices.

5.2. THE INPUT-OUTPUT TECHNOLOGY

5.2.1. Organic Light Emitting Diode (OLED), Micro-LED

Organic Laser Display uses the technology of Organic Light Emitting Semiconductor. OLED is a current-mode organic light-emitting device, which emits light through carrier injection and recombination. The luminous intensity is proportional to the current injected. Under the action of the electric field, the electron holes generated by the anode and the electrons generated by the cathode will move, inject into the electron hole transport layer and the electron transport layer respectively, and migrate to the luminescent layer. When they meet in the luminescent layer, they produce energy excitons, which excite the luminescent element and finally produce visible light [6].

The light-emitting processes of OLED devices can be divided into the injection of electrons and electron holes, the transport of electrons and electron holes, the recombination of electrons and electron holes.

The works in Luminescence are the following:

(1) Electron and hole injection. Electrons and holes in the cathode and anode will move to the luminous layer of the device driven by an applied driving voltage. In the process of moving to the luminous layer, the electrons and holes first need to pass through the cathode and electron injection layer and the anode and hole injection layer. The energy level barrier between the two layers moves to the electron transport layer and the hole transport layer of the device through the electron injection layer and the hole injection layer. The efficiency and lifetime of the device can be increased by the electron injection layer and the hole injection layer. The mechanism of electron injection in OLED devices is still being studied. The most commonly used mechanisms are tunneling effect and interface dipole mechanism.

(2) Electron and hole transport. Driven by an external driving voltage, the holes from the cathode and anode will move to the electron transport layer and the hole transport layer of the device respectively, and the electron transport layer and the hole transport layer will move the electrons and holes to the interface of the device's luminescent layer, respectively. At the same time, the electron transport layer and the hole transport layer will move to the interface of the device's luminescent layer separately. Holes from the anode and electrons from the cathode are blocked at the interface of the light emitting layer of the device, which makes the electrons and holes accumulate at the interface of the light emitting layer of the device.

(3) Recombination of electrons and holes. When the number of electrons and holes at the interface of the light emitting layer reaches a certain number, the electrons and holes will be recombined to produce excitons in the light emitting layer.

(4) Deexcitation light of exciton. Excitons generated in the luminescent layer will

activate the organic elements in the luminescent layer of devices, and then make the outermost electrons of organic elements transit from ground state to excited state. Because the electrons in the excited state are extremely unstable, they will transit to ground state. In the process of transition, energy will be released in the form of light. The light emission of the device is realized

The comparisons of three different kinds of OLED is shown in Table **5-1**.

Table 5-1. The comparisons of three different kinds of OLED.

	RGB-polymer emitters	Color filters White emitters	Color Changing Media (CCMS)
Advantages	Power efficient	Well-established technology	homogeneous aging Of emitter
	Lower production cost	No patterning of emitter necessary	More efficient than filters
	Mature ITO technology	homogeneous aging Of emitter	No pattering of emitter necessary
Disadvantages	Emitters have to be optimized separately	Power inefficient	ITO sputtering on CCMs
		ITO sputtering on filters	Stable blue emitter necessary
	Differential aging of emitters	Efficient white emitter necessary	Aging of CCMs

5.2.2. Human–computer Interaction

Intelligent man-machine interface is generally referred to as an intelligent interface. It aims to establish a harmonious man-machine interaction environment, to realize intelligence under harmonious conditions and achieve harmony with the purpose of intelligence. It makes the interaction between people and computers as natural and as convenient as the communication between people. It improves the friendliness of man-machine interaction. It is of great significance to improve people's application level of information systems and to promote the development of related industries [8].

1) Human-machine interface

Compared with the general human-machine interface, the meaning of Intelligent Interface includes:

• It is the intermediary between end users, domain experts, knowledge engineers

and knowledge sources.
- It includes computer hardware and software.
- Intelligence, that is to say, it can realize the same functions that middleman experts can accomplish.

2) User friendliness

It means that the system can adapt to all kinds of users and make it easy for new users who had not been trained or experienced to interact with the system.

With regarding to user-friendly standards, many information experts have proposed several different standards according to specific systems and conditions, such as Stibic's 25 input/output friendly standards and Kennedy's 12 human-computer communication rules.

This requires the application of cognitive theory to the study of user characteristics (behavior, psychology, needs). The system can actively guide interaction and feedback on current processing to users. Interactive language is concise, logical and helpful for memory. It can help correct users' errors. The error information given by the system should be clear, detailed, consistent and polite.

A. To produce a flexible, understandable and qualified output of results, and to make appropriate and understandable explanations to users.

B. Easy to use. Users can interrupt execution at any time (including normal interrupts and abnormal interrupts) and have the ability to recover. Users can also change their basic ways to enter and exit easily and directly.

C. Flexible can automatically adjust the interface for different types of users and requirements.

D. The learning environment can be provided according to users' needs.

3) Intelligent interface.

It should achieve the following main tasks:

A. Problem Description: Accept the user's problems and other information, and do some processing to produce a demand model or other internal form. Operational processing includes analyzing and understanding problems, understanding users' needs, identifying user-provided micro-concepts, and further explaining and expressing problems.

B. The expression and explanation of the system's answer: that is, the system's question, the answer to the user's question, the reasoning result and the interpretation of the result are reversed and transformed, and the user can understand the form of output to the user.

Session management: Session management can be divided into two parts. One part is the control of the overall structure of the conversation. This section describes how the session is controlled or executed, decides what questions to ask, and who (user or system) asks or answers when. The other part is to determine the conversational language.

Knowledge acquisition: acquire the knowledge needed by the system through the user's interactive dialogue.

(A) Problem description.

Problem description includes two tasks: one is natural language processing, grammatical and semantic analysis of input information; the other is processing the results of language understanding into a demand model (internal form) by using an expert's cognitive approach to deal with information problems.

To analyze and understand a problem is to clarify and explain the subject and context of the user's problem state or the user's abnormal knowledge state. Question description processing must apply many other elements, including user knowledge description, background information of professional field, available literature examples and description of professional literature. This is a complex multi-level problem. It is necessary to apply reasoning and knowledge acquisition techniques to derive and acquire unknown knowledge based on known descriptions.

Following are some relevant literature retrieval methods: Using the user knowledge model, the system can centralize to obtain a complete and accurate description of user knowledge. Sometimes it is necessary to extract background knowledge to verify user knowledge description. Professional information plays an important role in the problem description. The system can obtain relevant information from professional knowledge sources. If the source of expertise is inadequate, users can be asked to explain some aspects of their research to understand the subject of the problem, the scope of the problem and the various concepts involved. Sometimes it is difficult for users to explain the research topics, and they can obtain relevant information from users 'hobbies, places of residence and environment.

Sometimes topic-related descriptions occur in the initial conversation, and the system can visually inquire deeply about the information and obtain a general picture of the topic. The system can also ask users additional information to help develop the initial problem description. If there is no direct description related to the subject, the system can synthesize all the input information of the user and summarize the problem description. Based on the above processing conditions, it is required that the system can properly apply user model knowledge and domain knowledge to realize the five sub-functions of the problem description.

(B) The expression and interpretation of system answers are generally simple, such as the number of records returned from databases or text libraries, or the number of hits recorded. If faced with a wide range of problems, the system must be able to produce natural answers, that is, the output of the natural language processing system, to give natural explanations for inaccurate problems.

PROBLEMS

5-1 Briefly describe the definition, function, characteristics and performance parameters of bus.

5-2 Design a simple instrument based on CAN bus.

5-3 What is EPA and what are the main contents and characteristics of EPA system?

REFERENCES

[1]   "IIC bus" Available from:, https://en.wikipedia.org/wiki/

[2]   *I2C-bus specification and user manual.* NXP Semiconductors, 2014.

[3]   L. Fu, C.J. Deng, and L. Nie, *Theory, Design and Application of Intelligent Instruments,* Southwest Jiao tong University Press: China, Chengdu, .

[4]   H.W. Chen, "Development of Touch Screen Driver Based on S3C2410 under Linux", *Proceedings of the 2012 International Conference on Communication, Electronics and Automation Engineering,* 2013

[5]   "EPA bus" available from, https://en.wikipedia.org/wiki/

[6]   R. Singh, "Improving the contrast ratio of OLED displays: An analysis of various techniques", *Opt. Mater.,* vol. 34, no. 4, pp. 716-723, 2012.
[http://dx.doi.org/10.1016/j.optmat.2011.10.005]

[7]   F. Templier, *Overview of OLED Displays.* OLED Microdisplays, 2014.
[http://dx.doi.org/10.1002/9781119004745.ch2]

[8]   "HMI" available from:, https://en.wikipedia.org/wiki/

CHAPTER 6

Proteus and its Simulation Design Examples

Abstract: In this chapter, we will introduce two kinds of proteus designs, one is the foundational instrument hardware design example, and the other is its advance instrument design example.

Keywords: ADC, ARM, IoT, Key, LED, LCD, Proteus.

6.1. THE FOUNDATIONAL DESIGN EXAMPLE OF PROTEUS

6.1.1. The LED and Key Design [1]

The brightness of LED is better when the current is 8-20 mA. The resistance value in the circuit subtracts the saturated voltage drop of the diode from the current. The function of this program is to control IO, form a loop, and make the light on or off.

The key design is to make the IO port read in, scan the IO port, eliminate the jitter by delay, wait to release the key, and perform the key function. Its IO port is pulled up normally and grounded after keys putting down.

Figs. (**6-1 - 6-4**) are redrawn with reference [1].

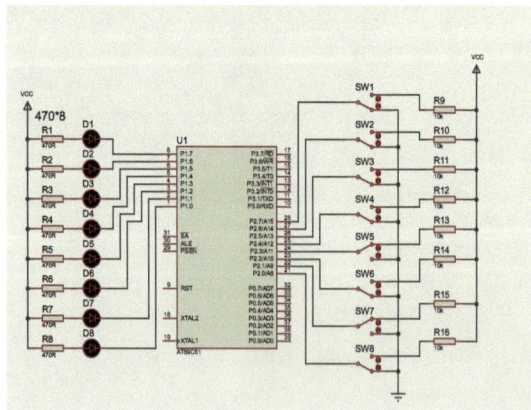

Fig. (6-1). The proteus example of led and key.

Example 6-1: Running-horse Lantern Design

Solution 6-1 [1]:

```c
#include <reg51.h>
#include <stdio.h>
unsigned int temp1;
void delay(unsigned int temp)
{
while(--temp); //delay
}
void main()
{
P1=255;//led is off
while(1)
{
P1=0XFE; //D8 light
temp1=35000;
delay(temp1);
P1=0XFD; //D7 light
temp1=35000;
delay(temp1);
P1=0XFB; //D6 light
temp1=35000;
delay(temp1);
P1=0XF7; //D5 light
```

temp1=35000;

delay(temp1);

P1=0XEF; //D4 light

temp1=35000;

delay(temp1);

P1=0XDF; //D3 light

temp1=35000;

delay(temp1);

P1=0XBF; //D2 light

temp1=35000;

delay(temp1);

P1=0X7F; //D1 light

temp1=35000;

delay(temp1);

}

}

Example 6-2: key control

Solution 6-2[1]:

ORG 000H

LJMP MAIN

ORG 0030H

MAIN: MOV SP,#7

MOV P1,#000H ;

MOV P2,#0FFH ;

LOOP: MOV A,P2 ;//read P2

MOV P1,A ; //light LED

MOV 20H,A ;

SCAN: MOV A,P2 ;

CJNE A,20H,LOOP ;

SJMP SCAN ;

END

6.1.2. The 7-seg LED and Matrix Key Design

Seven digital tubes have a common port of negative level of a common of positive level. Like the LED, the current is that each segment of the LED is bright, and the digital display is to make the corresponding LED bright according to the code.

Matrix keyboard has two working modes, one is line inversion, and the other is line scanning. In the mode of line inversion, the matrix rows constitute a working state of the preceding independent keys, one analyzes the column to press, and then invert; Then matrix columns constitute the preceding independent keys a working state, one assesses the line to press, and processes to get the keycode. In the mode of line scanning, it does not inverse, it determines which line of keys pressed using the scanning method. Case 6-2 gets a keycode using the inversion method.

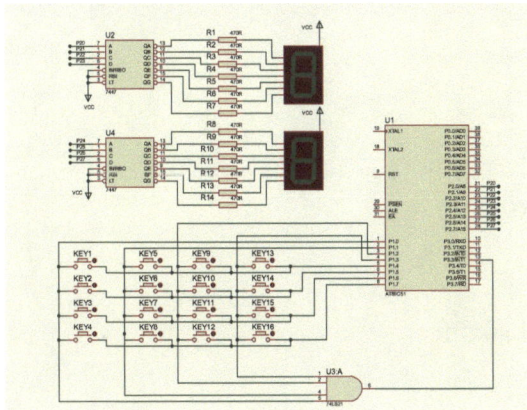

Fig. (6-2). The key matrix and 7-seg led example.

Example 6-3: The Matrix Key

Solution 6-3 [1]:

```c
#include <reg51.h>
#include <stdio.h>
#define byte unsigned char
void keyinterrupt() interrupt 2
{
int t;
byte keycode,scancode,flag=0xff;
t=5000;
while(t--);
if(INT1==1)
return;
EX1=0;
scancode=0xef;
while(scancode!=0xff)
{
P1=scancode;
keycode=P1;
if((keycode&0x0f)!=0x0f)
break;
scancode=(keycode<<1)|0x0f;
}
keycode=~keycode;
```

```
P2=keycode; //

P1=0X0F;

while(1)

{

if(INT1==1)

{

flag=~flag;

if(flag==0)

break;

}

t=10000;

while(t--);

}

EX1=1;

return;

}

void main(void)

{

IE=0;

EX1=1;

EA=1; //interrupt open

P2=0XFF;

P1=0X0F;

while(1)
```

```
{

}

}
```

6.1.3. The LCD and ADC Examples

1) LCD1602

LCD 1602 realizes LCD display by controlling its command word, writing LCD and reading LCD. The working is described as follows:

Owing to Instruction Coding Reference Data Manual, As shown in Example 6-2, the RS, RW and E lines control different addresses. It is mainly an address and subroutine of instruction class (instruction writing 'write_command'), judging busy address subroutine (instruction reading), writing data subroutine (data writing) of LCD, reading data from LCD subroutine (data reading). The write_command (0x80) writes data to the first line of liquid crystal, write_command (0xc0) writes data to the second line of liquid crystal. The routing of initialization includes the instruction writing, set data display mode (dot matrix type, display line number, data lines number), clear screen, set cursor place and open LCD display on.

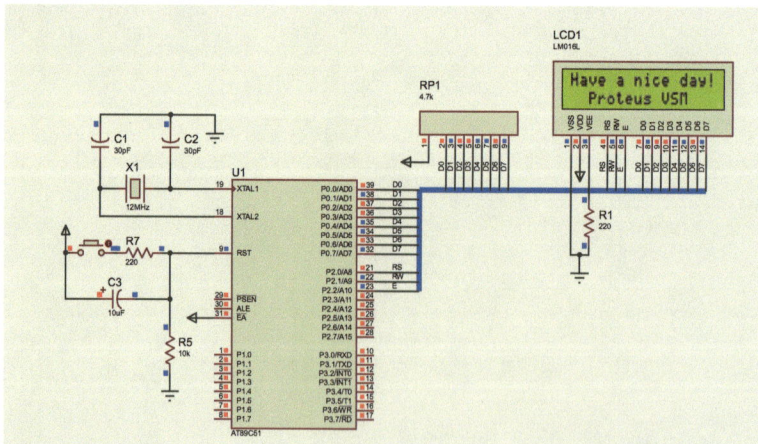

Fig. (6-3). The 1602 LCD example.

Example 6-3:

Solution 6-3 [1]:

```c
#include "ioAT89C51.h"
#include "intrinsics.h"
// Define P3 pins
#define DATA_BUS (P0)
#define RS (P2_bit.P2_0)
#define RW (P2_bit.P2_1)
#define E (P2_bit.P2_2)
// Define new types
typedef unsigned char uchar;
typedef unsigned int uint;
// Function Prototypes
void check_busy(void);
void write_command(uchar com);
void write_data(uchar data);
void LCD_init(void);
void string(uchar ad, uchar *s);
void lcd_test(void);
void delay(uint);
void main(void)
{ LCD_init();
while(1)
{ string(0x80,"Have a nice day!");
string(0xC0," Proteus VSM");
delay(100);
```

```
write_command(0x01);

delay(100);

}

}
```

/***

LCD1602 Driver mapped as IO peripheral

***/

```
// Delay

void delay(uint j)

{ uchar i = 60;

for(; j>0; j--)

{ while(--i);

i = 59;

while(--i);

i = 60;

}

}

// Test the Busy bit

void check_busy(void)

{ do

{ DATA_BUS = 0xff;

E = 0;

RS = 0;

RW = 1;
```

```
E = 1;

__no_operation();

} while(DATA_BUS & 0x80);

E = 0;

}
// Write a command
void write_command(uchar com)

{ check_busy();

E = 0;

RS = 0;

RW = 0;

DATA_BUS = com;

E = 1;

__no_operation();

E = 0;

delay(1);

}
// Write Data
void write_data(uchar data)

{ check_busy();

E = 0;

RS = 1;

RW = 0;

DATA_BUS = data;
```

```
E = 1;

__no_operation();

E = 0;

delay(1);

}
// Initialize LCD controller
void LCD_init(void)
{ write_command(0x38); // 8-bits, 2 lines, 7x5 dots
write_command(0x0C); // no cursor, no blink, enable display
write_command(0x06); // auto-increment on
write_command(0x01); // clear screen
delay(1);

}
// Display a string
void string(uchar ad, uchar *s)
{ write_command(ad);
while(*s>0)
{ write_data(*s++);
delay(100);

}

}
```

1) AD1674 example

The principle is explained in chapter 2: first, start a conversion with A0 and CS low; then the conversion takes place on rising CE edge; and then wait until we have completed a conversion, in the end read the measurement data and output the measurement data.

The Figs. (**6-4-6-6**) are redrawn with reference [2].

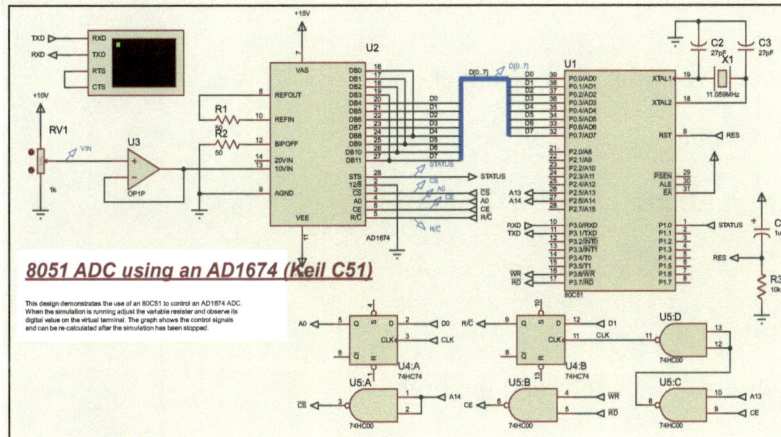

Fig. (6-4). The ad1674 example.

Example 6-4: ad conversion example

Solution 6-4 [2]:

#include <reg51.h>

#include <INTRINS.H>

#include <STDIO.H>

// define P1.0 to check STATUS.

sbit STATUS = P1^0;

unsigned char xdata CTRL _at_ 0x2FFF;

unsigned char xdata ADSEL _at_ 0x4FFF;

unsigned char hByte;

unsigned char lByte;

void adc_Convert (void)

{ // Start a conversion with A0 and A/C low.

// The conversion takes place on rising CE edge.

CTRL = 0x00;

ADSEL = 0x00;

// Wait until we have completed a conversion .

while(STATUS==1);

// Set R/C with A0 low and read the low byte.

CTRL = 0x02;

hByte = ADSEL;

// Set R/C with A0 high and read the high.

CTRL = 0x03;

lByte = ADSEL;

}

void main(void)

{ unsigned int delay, MSB, LSB, adc_Res;

// Initialize serial interface

SCON = 0xDA; // SCON: mode 1, 8-bit UART, enable rcvr */

TMOD |= 0x20; // TMOD: timer 1, mode 2, 8-bit reload */

TH1 = 0xFD; // TH1: reload value for 1200 baud @ 12MHz */

TR1 = 1; // TR1: timer 1 run */

TI = 1; // TI: set TI to send first char of UART */

while(1)

{ adc_Convert();

MSB=(unsigned int)(hByte << 4);

LSB=(unsigned int)(lByte >> 4);

// adc_Res now has the converted data with 12-bit resolution.

adc_Res = MSB + LSB;

// Send adc results to the serial interface

printf("ADC READINGS: %03Xh\n", adc_Res);

// simple delay - it is mcu clock dependent !

for (delay=0; delay<10000; delay++)

;

}

}

2) Graphics LCD and LPC2138

Graphics LCD is similar as LCD1602.LPC2138 should select PLL clock, and set VIC in its Target.c

Fig. (6-5). The LPC2138 with graphics LCD example.

Example 6-5: the LPC2138 with graphics LCD

Solution 6-5 [2]:

```
/***************************************************************
***********
```

```
* FileName:LCDTEST.C

* Function:Display a picture

*****************************************************************
**********/

#include "Config.h"

#include "Target.h"

#include "T6963C.h"

/****************************************************************
**********

* Function Name: DelayNS()

* Function Description: Software delay

* Input parameter: dly delay

* Output parameter: none

*****************************************************************
**********/

void DelayNS(uint32 dly)
{ uint32 i;
for(; dly>0; dly--)
for(i=0; i<10000; i++);
}
void init_lcd (void)
{ IO0CLR=rst;
IO0SET=rst;
delay1(50);
IO0CLR=ce;
```

```
IO0SET=wr;

IO0SET=rd;

wr_xd(addr_w,0x40); // Set text display buffer base address

wr_xd(addr_t,0x42); // Set graphics display buffer base address

wr_td(width,0x00,0x41); // Set text display width

wr_td(width,0x00,0x43); // Set graphics display width

wr_comm(0x81); // Set display mode: text xor graphics

wr_td(0x56,0x00,0x22); // Set CGRAM offset address

wr_comm(0x9c); // Set text and graphic display on

}
// Main Program

void main ()
{ CCR_bit.CLKSRC = 1; // Use external clock

PINSEL0 = 0x00000000; // All GPIO

PINSEL1 = 0x00000000; // All GPIO

IO0DIR = 0xffffffff;

init_lcd (); // Initialize LCD

while(1)
{ clrram(); // Clear the screen

disp_img(0,16,64,nBitmapDot); // Display a bitmap

//disp_hz(4,4,"Guangzhou Fengbiao");

disp_zf(0,10, "Text:"); // Display a string

disp_zf(1,12, "LABCENTER"); // Display a string

disp_zf(1,14, "Proteus VSM");
```

DelayNS(150);

clrram();

disp_dz(0xcc,0x33); // Display a matrix

DelayNS(100);

}

}

/***
****

- FB-EDU-PARM-LPC2138 Target Initial file

****
***********************************/

#include "Target.h"

#include "Config.h"

/***
****

** Function Name: IRQ_Exception

** Function Desc: Interrupt exception handler

**

** Input: none

**

** Output: none

** __ICFEDIT_region_ROM_start__

** Global var: none

** Called module: none

**

```
*****************************************************************
*************************************/

__irq void IRQ_Exception(void)

{

while(1); // You should replace the code here if you are using IRQ

}

/***************************************************************
***************************************

** Function Name: FIQ_Exception

** Function Desc: Fast Interrupt exception handler

**

** Input: none

**

** Output: none

**

** Global var: none

** Called module: none

***************************************************************
*************************************/

void FIQ_Exception(void)

{

while(1); // You should replace the code here if you are using FIQ

}

/***************************************************************
***************************************

** Function Name: TargetInit
```

** Function Desc: Target board initialisation routine

**

** Input: none

**

** Output: none

**

** Global var: none

** Called module: none

****

***************************************/

void TargetInit(void)

{

/* Add your code here */

}

/***

****

** Function Name: TargetResetInit

** Function Desc: Initialisation after reset

**

** Input: none

**

** Output: none

**

** Global var: none

** Called module: none

```
*****************************************************************
************************************/

void TargetResetInit(void)

{

#ifndef NDEBUG

MEMMAP = 0x2; //remap

#else

MEMMAP = 0x1; //remap

#endif

/* Set up System clock */

PLLCON = 1;

#if (Fpclk / (Fcclk / 4)) == 1

VPBDIV = 0;

#endif

#if (Fpclk / (Fcclk / 4)) == 2

VPBDIV = 2;

#endif

#if (Fpclk / (Fcclk / 4)) == 4

VPBDIV = 1;

#endif

#if (Fcco / Fcclk) == 2

PLLCFG = ((Fcclk / Fosc) - 1) | (0 << 5);

#endif

#if (Fcco / Fcclk) == 4
```

```
PLLCFG = ((Fcclk / Fosc) - 1) | (1 << 5);

#endif

#if (Fcco / Fcclk) == 8

PLLCFG = ((Fcclk / Fosc) - 1) | (2 << 5);

#endif

#if (Fcco / Fcclk) == 16

PLLCFG = ((Fcclk / Fosc) - 1) | (3 << 5);

#endif

PLLFEED = 0xaa;

PLLFEED = 0x55;

while((PLLSTAT & (1 << 10)) == 0);

PLLCON = 3;

PLLFEED = 0xaa;

PLLFEED = 0x55;

/* Set up memory accelerate module */

MAMCR = 0;

#if Fcclk < 20000000

MAMTIM = 1;

#else

#if Fcclk < 40000000

MAMTIM = 2;

#else

MAMTIM = 3;

#endif
```

#endif

MAMCR = 2;

/* Init VIC */

VICIntEnClear = 0xffffffff;

VICVectAddr = 0;

VICIntSelect = 0;

/* Add your additional init code here */

}

6.2. THE ADVANCED DESIGN EXAMPLE OF PROTEUS

6.2.1. The IoT Design by Proteus [3]

The IoT and its associated technologies are currently transforming the world of purely digital information systems into Cyber-physical Systems (CPS). The interaction with the physical world requires feedback loops and flexible service composition to compensate for errors, which makes the selection of IoT services highly context dependent.

To meet the needs of modern meteorological detection, automatic weather station of the Internet of Things is developed. Local area wireless sensor networks are composed of temperature, humidity, wind, pressure, precipitation and other meteorological sensor nodes, routers, and the coordinator.

Wide area networks are also composed of the data communicator and remote computer through 3G, GPRS and the Internet. Data are received, analyzed and stored in real-time network databases by the remote computer after the quality control. Meteorological data management and distribution are implemented, and the data can be consulted through the user's mobile phone or Internet computer. The Figs. (6-6, 6-7) are redrawn with [3].

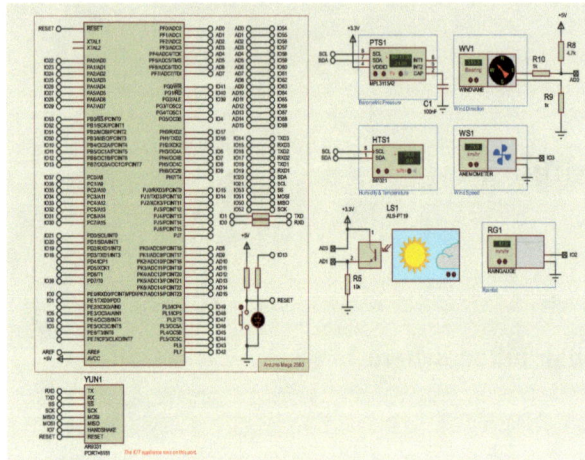

Fig. (6-6). The proteus schematic of a weather station.

6.2.1. The IOT and YUN Design by Proteus

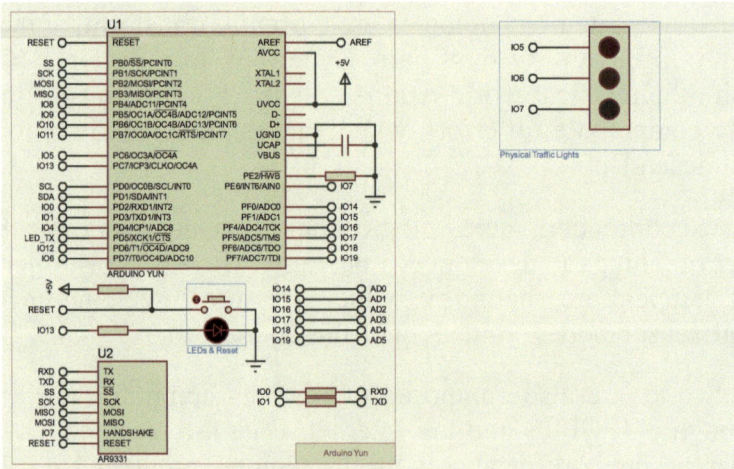

Fig. (6-7). The Proteus Schematic of Traffic Light.

The Python code for traffic light is presented in the following [3]:

VFP Server

import os, sys, stat, select, fcntl

import socket, select, struct

```
import time, calendar

import re, threading

HOST = '' # Symbolic name meaning all available interfaces

PORT = 8080 # Arbitrary non-privileged port

BUFFERSIZE = 4096

INITD_SCRIPT = "/etc/init.d/iotbuilder-publish"

AVAHI_DAEMON = "/etc/init.d/avahi-daemon"

AVAHI_SERVICE = "/vsm/iotbuilder.service"

PANEL_SVG = "/vsm/panel.svg"

LOG_NAME = "/tmp/server.log"

KEEPALIVE_INTERVAL = 10

KEEPALIVE_TIMEOUT = 30

global applianceName

global clientAddress

global statusSocket

global statusBuffer

global sessionState

global sessionHistory

def processRequest (conn, addr):

global applianceName

global clientAddress

global statusSocket

global statusBuffer

firstLine = ""
```

```
currentLine = ""

lines = [];

done = False

#writelog("Receiving...")

conn.settimeout(5.0)

connfile=conn.makefile('r', 1)

while not done:

try:

data = connfile.readline()

currentLine = data.split('\r')[0]

currentLine = currentLine.split('\n')[0]

if currentLine != "":

lines.append(currentLine)

elif firstLine == "":

firstLine = lines[0]

done = firstLine.startswith('GET')

body = len(lines)

else:

done = True

except socket.timeout:

writelog("Socket Timeout")

conn.close()

return

except:
```

```
writelog("Socket Error")

conn.close()

return

conn.settimeout(None)

if firstLine != "":

writelog(firstLine)

action = firstLine.split(' ')[0];

if action=='GET':

filename = firstLine.split(' ') [1];

if filename == '/':

#Process a page reload.

if clientAddress == None or clientAddress == addr:

# If this is a new connection, or a reload from the same client then this is
processed normally:

filename = 'panel.htm'

writelog("Client Address:"+str(addr))

print "RELOAD:*"

# Close any previous status connection and launch the status timeout. Anything
that doesn't establish a

# status connection within the status timeout period will thus be disconnected.

clientAddress = addr

statusBuffer = "

closeStatus()

else:

# If an attempt is made to connection from another client without first closing the
```

other, then send a 403:

```
conn.sendall("HTTP/1.1 403 FORBIDDEN\n")

conn.sendall("Content-Type: text/html\n\n")

conn.sendall("<html>")

conn.sendall("<head><title>Appliance In Use</title></head>")

conn.sendall("<body><h1>")

conn.sendall("The '"+applianceName+"' is under the control of another client ["+str(clientAddress)+"].\n")

conn.sendall("</h1></body>")

conn.sendall("</html>")

conn.close()

writelog("Rejected connection from " +str(addr))

return

elif filename == '/status':

#Process a status request.

if clientAddress == addr:

#We send a response header but then keep the connection

#open until we receive something to send to it from the Arduino

conn.sendall("HTTP/1.1 200 OK\n")

conn.sendall("Content-Type: text/plain\n")

conn.sendall("Connection: close\n")

conn.sendall("\n")

if statusBuffer == ":

openStatus(conn)
```

```
else:

conn.sendall(statusBuffer)

conn.close()

statusBuffer = "

else:

# Another client is trying to use the appliance:

writelog("Rejected client:"+str(addr))

conn.sendall("HTTP/1.1 403 FORBIDDEN\n")

conn.sendall("Content-Type: text/plain\n")

conn.sendall("Connection: close\n")

conn.sendall("\n\n")

conn.close();

return

elif filename == '/session':

#Process a session state request.

openStatus(conn)

sendState(conn)

closeStatus()

return

elif filename.startswith('/'):

filename = filename[1:]

sendFile(conn, filename)

elif action=='POST' or action=='PUT':

# Process messages from the client/browser.
```

POST messages are passed on to the AVR, PUT messages merely update the session state.

```
conn.sendall("HTTP/1.1 200 OK\n")

conn.sendall("Content-Type: text/plain\n")

conn.sendall("Connection: close\n")

conn.sendall("\n")

for i in range(body, len(lines)):

currentLine = lines[i]

writelog(currentLine)

if len(currentLine) > 0:

if action=='POST':

writelog("EVENT:"+currentLine)

print "EVENT:"+currentLine+"*"

else:

writelog("RECORD:"+currentLine)

saveState(currentLine)

conn.close()

#Opens a requested file

def sendFile (conn, filename):

#Reccords server acknowledgement

writelog("Sending file '"+filename+"'")

#Trys to send the file

try:

file = open(filename, "rb")
```

```
except:
conn.sendall("HTTP/1.1 404 Not Found\n")
conn.sendall("Content-Type: text/html\n\n")
conn.sendall("<html>")
conn.sendall("<head><title> 404 NOT FOUND </title></head>")
conn.sendall("<body><h1>")
conn.sendall(filename+" - file not found")
conn.sendall("</h1></body>")
conn.sendall("</html>")
writelog("File failed to send" + "\n")
else:
conn.sendall("HTTP/1.1 200 OK\n")
conn.sendall("Content-Type: " + fileType(filename) + "\n\n")
conn.sendall(file.read()) # TBD send in buffersize chunks?
writelog("Transfer successful")
#Send the session state (state variables of all controls)
def sendState(conn):
conn.sendall("HTTP/1.1 200 OK\n")
conn.sendall("Content-Type: text/plain\n")
conn.sendall("Connection: close\n")
conn.sendall("\n")
writelog("Sending session state")
for key in sessionState:
if not key.startswith('$.record'):
```

```
conn.sendall(key+'='+sessionState[key]+'\n')

for key in sessionHistory:

conn.sendall(sessionHistory[key]);

def fileType (filename):

extn = filename.split('.') [1];

if extn == "png":

filetype = "image/png"

elif extn == "gif":

filetype = "image/gif"

elif extn == "jpeg":

filetype = "image/jpeg"

elif extn == "svg":

filetype = "image/svg+xml"

elif extn == "js":

filetype = "application/javascript"

elif extn == "htm" or extn == "html":

filetype = "text/html"

elif extn == "txt":

filetype = "text/html"

elif extn == "css":

filetype = "text/css"

else:

filetype = "application/octet-stream"

return filetype
```

```python
def pollStatus ():

global statusSocket

global statusBuffer

global keepaliveTimer

data = ""

# Read pending data from the AVR side

while sys.stdin in select.select([sys.stdin], [], [], 0)[0]:

line = sys.stdin.readline()[:-1]

writelog("Status: "+line)

if saveState(line):

data += line+'\n'

# If the keepalive timer has expired we need to send a keep alive packet

if len(data) == 0 and keepaliveTimer != None and not keepaliveTimer.is_alive():

writelog('Sending Keepalive');

data = "$.keepalive\n"

keepaliveTimer = None

# If we have a status connection then use it to send the data,

# otherwise we park it in a buffer unless the client has gone away.

if len(data) != 0:

if statusSocket != None:

try:

statusSocket.sendall(data)

except IOError:

pass
```

```
closeStatus()

elif statusTimeout != None:

statusBuffer += data

else:

statusBuffer = "

return
```

This is called when the client opens a reverse AJAX connection and we have nothing to send.

```
def openStatus (sock):

global statusSocket;

global keepaliveTimer

global statusTimeout

if statusSocket == None and statusTimeout == None:

# New Connection

writelog("Client connected:"+str(sock.getpeername()[0]))

if statusSocket != None:

statusSocket.close()

statusSocket = sock

keepaliveTimer = threading.Timer(KEEPALIVE_INTERVAL, keepaliveDue)

keepaliveTimer.start()

if statusTimeout != None:

statusTimeout.cancel()

statusTimeout = None
```

This is called to close status connection because we want the client to process the information

that we have written to it. At this point we start the a timeout timer, because we except to receive

another status connection within this time frame.

def closeStatus ():

global statusSocket;

global keepaliveTimer

global statusTimeout

if statusSocket != None:

statusSocket.close()

statusSocket = None

if keepaliveTimer != None:

keepaliveTimer.cancel()

keepaliveTimer = None

if statusTimeout != None:

statusTimeout.cancel()

statusTimeout = threading.Timer(KEEPALIVE_TIMEOUT, lostStatus)

statusTimeout.start()

This is triggered if/when the keepalive timer triggers.

It will cause the next call to pollStatus() to post a keepalive message and close the status pipe

at which point the server should request a new one.

def keepaliveDue():

pass

If the statusTimeout timer triggers then we can assume that we have lost contact with the client.

```python
def lostStatus():

global keepaliveTimer

global statusTimeout

global clientAddress

writelog("Client disconnected")

print "DISCONNECT:*"

clientAddress = None

if keepaliveTimer != None:

keepaliveTimer.cancel()

keepaliveTimer = None

if statusTimeout != None:

statusTimeout.cancel()

statusTimeout = None

# Update the recorded state data.

# Returns true is there is a state change which should be passed to the browser.

def saveState(msg):

global sessionState

global sessionHistory

m = re.match("([$a-zA-Z_][a-zA-Z0-9_\\.]*)\\s*=\\s*([^\\r\\n]*)", msg)

dirty = False

if m != None and len(m.groups()) == 2:

# State assignment and history functions:

key = m.group(1)

value = m.group(2)
```

```python
if key.startswith("$.create"):

dirty = True;

elif key.startswith("$.record"):

# Start recording for control with id=value

id = value.replace("", ")

sessionState["$.record."+id] = "1"

elif key.startswith("$.stop"):

# Stop recording for control with id=value

id = value.replace("", ")

sessionState["$.record."+id] = "0"

elif key.startswith("$.erase"):

# Erase recorded history for control with id=value

id = value.replace("", ")

sessionHistory.pop(id, None)

elif not key in sessionState or sessionState[key] != value:

# Normal state assignment

sessionState[key] = value

dirty = True

else:

# Ordinary JS method call of form <id>.<method> (<args>)

id = msg.split('.')[0]

if sessionState.get("$.record."+id, '0') == '1':

h = sessionHistory.get(id, ")+msg+'\n';

sessionHistory[id] = h
```

```
dirty = True;

return dirty

def writelog(msg):

if os.path.exists(LOG_NAME):

log = os.open(LOG_NAME, os.O_WRONLY|os.O_APPEND)

os.write(log, msg+'\n');

os.close(log)

def get_ip_address(ifname):

s = socket.socket(socket.AF_INET, socket.SOCK_DGRAM)

return socket.inet_ntoa(fcntl.ioctl(

s.fileno(),

0x8915, # SIOCGIFADDR

struct.pack('256s', ifname[:15])

)[20:24])

def configure (title):

# Create the IotBuilder publishing script within /etc/init.d

writelog("Creating "+INITD_SCRIPT);

file = os.open(INITD_SCRIPT, os.O_WRONLY|os.O_CREAT|os.O_TRUNC)

os.write(file, "#!/bin/sh /etc/rc.common\n")

os.write(file, "# Publish/Unpublish the IotBuilder service file.\n")

os.write(file, "STOP=99\n")

os.write(file, "start() {\n cp /vsm/iotbuilder.service /etc/avahi/services\n}\n")

os.write(file, "stop() {\n rm /etc/avahi/services/iotbuilder.service\n}\n")

os.close(file)
```

```
os.chmod(INITD_SCRIPT, 0o777) # mark as executable

os.system(INITD_SCRIPT + " enable")

# Create the avahi-service descriptor file

writelog("Creating "+AVAHI_SERVICE);

file                    =                    os.open(AVAHI_SERVICE,
os.O_WRONLY|os.O_CREAT|os.O_TRUNC)

os.write(file, "<?xml version='1.0' standalone='no'?>\n")

os.write(file, "<!DOCTYPE service-group SYSTEM 'avahi-service.dtd'>\n")

os.write(file, "<service-group>\n")

os.write(file, " <name replace-wildcards='yes'>"+title+" on %h</name>\n")

os.write(file, " <service>\n")

os.write(file, " <type>_vfpserver._tcp</type>\n")

os.write(file, " <port>"+str(PORT)+"</port>\n") ## Could choose arbitrary/free
port here

os.write(file, " </service>\n")

os.write(file, "</service-group>\n")

os.chmod(AVAHI_SERVICE, 0o644) # mark as non-executable

os.close(file);

# Publish the Service:

os.system(INITD_SCRIPT + " start")

os.system(AVAHI_DAEMON + " restart")

### Main Logic

# Set the CWD to where we are running:

cwd = os.path.dirname(sys.argv[0])

if cwd != "":
```

```
os.chdir(cwd)

# Set the PORT if specified on the command line:

if len(sys.argv) >= 2:

PORT = int(sys.argv [1])

# Extract the project title from panel.svg

# This should work using the XML library but the parser library is missing on the
Yun Board.

from HTMLParser import HTMLParser

class SvgParser(HTMLParser):

def __init__(self):

HTMLParser.__init__(self)

self.title = 'Virtual Front Panel'

def handle_starttag(self, tag, attrs):

if (tag == "svg"):

for attr in attrs:

if attr[0] == 'vfp:title':

self.title = attr [1]

# instantiate the parser and fed it some HTML

parser = SvgParser()

parser.feed(open(PANEL_SVG).read())

applianceName = parser.title

writelog("Configuring '"+parser.title+"'")

#Create, bind and listen on the socket:

sock = socket.socket(socket.AF_INET, socket.SOCK_STREAM)
```

```
sock.setsockopt(socket.SOL_SOCKET, socket.SO_REUSEADDR, 1)

sock.bind((HOST, PORT))

sock.listen(1)

#Perform configuration operations:

configure(parser.title)

clientAddress = None

statusSocket = None

statusBuffer = ""

sessionState = {}

sessionHistory = {}

keepaliveTimer = None

statusTimeout = None

# Have we got wlan0:

try:

ipAddr = get_ip_address('wlan0')

writelog("Listening on wlan0:"+ipAddr+":"+str(PORT))

except:

writelog("No IP for wlan0")

# Have we got eth0:

try:

ipAddr = get_ip_address('eth0')

writelog("Listening on eth0:"+ipAddr+":"+str(PORT))

except:

writelog("No IP for eth0")
```

```
# Have we got eth1:

try:

ipAddr = get_ip_address('eth1')

writelog("Listening on eth1:"+ipAddr+":"+str(PORT))

except:

writelog("No IP for eth1")

# Start of 2017 - this will prevent us from emitting the time until the Linux

# side has got a proper time off an internet time server.

next = 1483228800

while True:

# Get the current time and convert non UTC value for use by the AVR side

now = calendar.timegm(time.localtime())

if now > next:

print "TIME:"+str(now)+"*"

next = now+3600;

sock.setblocking(False);

try:

conn, addr = sock.accept()

processRequest(conn, addr[0]);

except socket.error:

#No connection

pass

pollStatus()
```

PROBLEM

6-1 realize all the examples using Proteus software.

REFERENCES

[1]   X.L. Li, *Proteus based 8051 MCU case tutorial.* Publishing House of Electronics Industry: Beijing, China, 2014.

[2]   "IoT builder", available from https://www.labcenter.com/iotbuilder [2019.2.1].

## CHAPTER 7

Arduino and MATLAB in Intelligent Instrument

Abstract: In this chapter, we will introduce two interesting domains in intelligent instrument: MATLAB and Arduino. As they easily combine with the theory and technology, some MATLAB design examples or demos are introduced including the modern signal process, machine vision, and so on. Similar to MATLAB, as the opensource hardware is easy to combine with the device, the Arduino SLAM, PHM and other intelligent instrument applications are introduced.

Keywords: Arduino, Computer Vision System, Matlab, Modern signal process, Predictive maintenance, PHM, SLAM.

7.1. MATLAB IN INTELLIGENT INSTRUMENT

MATLAB is a programming language and platform that is most suitable for solving problems that can be expressed by matrix and array mathematics directly. It is one of the most developed fields of modern intelligent instrument. We introduce some new toolkits and examples used in industrial instruments.

7.1.1. Predictive Maintenance Toolbox

This toolbox can obtain features using data-based and model-based techniques, including statistical, spectral, and time-series analysis. For example, it can be applied to monitor the health of rotating machines such as bearings and gearboxes using vibration data.

The process to train a machine learning model is in four steps: the first step is to detect anomalies, the second step is to classify different types of faults, the third step is to estimate the remaining useful life (RUL) of machine, and the final step is to deploy the algorithm and integrate into the online system for machine monitoring and maintenance. This process is shown in Fig. (7-1) [1].

ACQUIRE DATA	→	PREPROCESS DATA	→	IDENTIFY CONDITION INDICATORS	→	TRAIN MODEL	→	DEPLOY AND INTEGRATE

Fig. (7-1). The process of machine monitoring and maintenance.

1. The first step is to collect a large set of sensor data representing the healthy and faulty operation. It needs the data set in a different condition, or as an alternative, it needs a mathematical model. The combination of synthetic and sensor data is useful to develop a predictive maintenance algorithm.
2. The next step is to preprocess the data to convert it to a form from which condition indicators can be easily extracted. Preprocessing includes techniques such as noise, outlier, and missing value removal.
3. The next step is to identify condition indicators, features whose behavior changes in a predictable way as the system degrades. These features are used to discriminate between healthy and faulty operation. (Fig. **7-2**) shows condition indicators using both signal-based and model-based approaches. Some examples include:time-frequency moments, phase-space reconstructions, and so on [1].
4. In the next step, use the extracted features to train machine learning models to do several things: Detect anomalies (track changes in your system to determine the presence of anomalies); Detect different types of faults through classification; Predict the transition from healthy state and failure, (Finding a model that captures the relationship between the extracted features and the degradation path) it will help you estimate how much time there is until failure (remaining useful life) and when you should schedule maintenance. This is shown in Fig. (**7-3**).
5. After developing your algorithm, you can get it up and running by deploying it on the cloud or your edge device. A cloud implementation can be useful when you are also gathering and storing large amounts of data on the cloud. Alternatively, the algorithm can run on embedded devices that are closer to the actual equipment. This may be the case if an internet connection is not available. A third option is to use a combination of the two. If you have a large amount of data, and if there are limits on how much data you can transmit, you can perform the preprocessing and feature extraction steps on your edge device and then send only the extracted features to your prediction model that runs on the cloud. Data scientists often refer to the ability to share and explain results as model interpretability. An easily interpretable model has the following features:
• A small number of features that typically are created from some physical understanding of the system.
• A transparent decision-making process Interpretability is important for applications when you need to:

1. Prove that your model complies with government or industry standards
2. Explain factors that contributed to a diagnosis
3. Show the absence of bias in decision-making

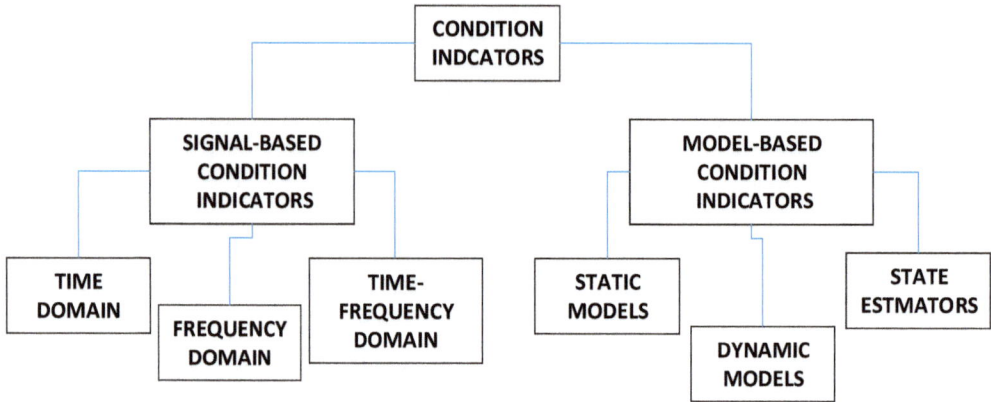

Fig. (7-2). The process of designing condition indicators.

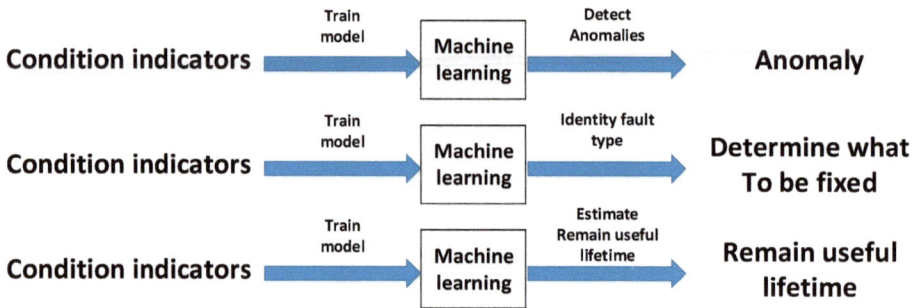

Fig. (7-3). Three ways to estimate the remaining useful life.

Example 7-1: Using Simulink to Generate Fault Data

Solution 7-1:

```
mdl = 'pdmTransmissionCasing';

open_system(mdl)

open_system([mdl '/Casing'])

open_system([mdl '/Vibration sensor with drift'])

open_system([mdl '/Shaft'])
```

```matlab
open_system([mdl,'/Shaft/Output Shaft'])

toothFaultArray = -2:2/10:0; % Tooth fault gain values

sensorDriftArray = -1:0.5:1; % Sensor drift offset values

shaftWearArray = [0 -1]; % Variants available for drive shaft conditions

% Create an n-dimensional array with combinations of all values

[toothFaultValues,sensorDriftValues,shaftWearValues] = ...

ndgrid(toothFaultArray,sensorDriftArray,shaftWearArray);

for ct = numel(toothFaultValues):-1:1

% Create a Simulink.SimulationInput for each combination of values

siminput = Simulink.SimulationInput(mdl);

% Modify model parameters

siminput = setVariable(siminput,'ToothFaultGain',toothFaultValues(ct));

siminput = setVariable(siminput,'SDrift',sensorDriftValues(ct));

siminput = setVariable(siminput,'ShaftWear',shaftWearValues(ct));

% Collect the simulation input in an array

gridSimulationInput(ct) = siminput;

end

rng('default'); % Reset random seed for reproducibility

toothFaultArray = [0 -rand(1,6)]; % Tooth fault gain values

sensorDriftArray = [0 randn(1,6)/8]; % Sensor drift offset values

shaftWearArray = [0 -1]; % Variants available for drive shaft conditions

%Create an n-dimensional array with combinations of all values

[toothFaultValues,sensorDriftValues,shaftWearValues] = ...

ndgrid(toothFaultArray,sensorDriftArray,shaftWearArray);
```

```
for ct=numel(toothFaultValues):-1:1
% Create a Simulink.SimulationInput for each combination of values
siminput = Simulink.SimulationInput(mdl);

% Modify model parameters
siminput = setVariable(siminput,'ToothFaultGain',toothFaultValues(ct));
siminput = setVariable(siminput,'SDrift',sensorDriftValues(ct));
siminput = setVariable(siminput,'ShaftWear',shaftWearValues(ct));

% Collect the simulation input in an array
randomSimulationInput(ct) = siminput;
end rng('default'); % Reset random seed for reproducibility
toothFaultArray = [0 -rand(1,6)]; % Tooth fault gain values
sensorDriftArray = [0 randn(1,6)/8]; % Sensor drift offset values
shaftWearArray = [0 -1]; % Variants available for drive shaft conditions
%Create an n-dimensional array with combinations of all values
[toothFaultValues,sensorDriftValues,shaftWearValues] = ...
ndgrid(toothFaultArray,sensorDriftArray,shaftWearArray);
for ct=numel(toothFaultValues):-1:1
% Create a Simulink.SimulationInput for each combination of values
siminput = Simulink.SimulationInput(mdl);

% Modify model parameters
siminput = setVariable(siminput,'ToothFaultGain',toothFaultValues(ct));
siminput = setVariable(siminput,'SDrift',sensorDriftValues(ct));
siminput = setVariable(siminput,'ShaftWear',shaftWearValues(ct));
```

% Collect the simulation input in an array

randomSimulationInput(ct) = siminput;

end

% Run the simulations and create an ensemble to manage the simulation results

if ~exist(fullfile(pwd,'Data'),'dir')

mkdir(fullfile(pwd,'Data')) % Create directory to store results

end

runAll = true;

if runAll

[ok,e]   = generateSimulationEnsemble([gridSimulationInput,randomSimulation Input], ...

fullfile(pwd,'Data'),'UseParallel', true);

else

[ok,e]   =   generateSimulationEnsemble(gridSimulationInput(1:10), fullfile(pwd,'Data')); %#ok<*UNRCH>

end

ens = simulationEnsembleDatastore(fullfile(pwd,'Data'));

ens.SelectedVariables = ["Vibration" "Tacho" "SimulationInput"];

data = read(ens)

vibration = data.Vibration{1};

plot(vibration.Time,vibration.Data)

title('Vibration')

ylabel('Acceleration')

idx = vibration.Time >= seconds(10);

vibration = vibration(idx,:);

```
vibration.Time = vibration.Time - vibration.Time(1);

tacho = data.Tacho{1};

idx = tacho.Time >= seconds(10);

tacho = tacho(idx,:);

plot(tacho.Time,tacho.Data)

title('Tacho pulses')

legend('Drive shaft','Load shaft') % Load shaft rotates more slowly than drive
shaft

idx = diff(tacho.Data(:,2)) > 0.5;

tachoPulses = tacho.Time(find(idx)+1)-tacho.Time(1)

vars = data.SimulationInput{1}.Variables;

idx = strcmp({vars.Name},'SDrift');

if any(idx)

sF = abs(vars(idx).Value) > 0.01; % Small drift values are not faults

else

sF = false;

end

idx = strcmp({vars.Name},'ShaftWear');

if any(idx)

sV = vars(idx).Value < 0;

else

sV = false;

end

if any(idx)
```

```
idx = strcmp({vars.Name},'ToothFaultGain');

sT = abs(vars(idx).Value) < 0.1; % Small tooth fault values are not faults

else

sT = false

end

faultCode = sF + 2*sV + 4*sT; % A fault code to capture different fault
conditions

sdata = table({vibration},{tachoPulses},sF,sV,sT,faultCode, ...

'VariableNames',{'Vibration','TachoPulses','SensorDrift','ShaftWear','ToothFault','
FaultCode'})

ens.DataVariables = [ens.DataVariables; "TachoPulses"];

ens.ConditionVariables                                              =
["SensorDrift","ShaftWear","ToothFault","FaultCode"];

reset(ens)

runLocal = false;

if runLocal

% Process each member in the ensemble

while has data(ens)

data = read(ens);

addData = prepareData(data);

writeToLastMemberRead(ens,addData)

end

else

% Split the ensemble into partitions and process each partition in parallel

n = numpartitions(ens,gcp);
```

```
parfor ct = 1:n

subens = partition(ens,n,ct);

while hasdata(subens)

data = read(subens);

addData = prepareData(data);

writeToLastMemberRead(subens,addData)

end

end

end

reset(ens)

ens.SelectedVariables = "Vibration";

figure,

ct = 1;

while hasdata(ens)

data = read(ens);

if mod(ct,10) == 0

vibration = data.Vibration{1};

plot(vibration.Time,vibration.Data)

hold on

end

ct = ct + 1;

end

hold off

title('Vibration signals')
```

```
ylabel('Acceleration')

featureVariables = analyzeData('GetFeatureNames');

ens.DataVariables = [ens.DataVariables; featureVariables];

ens.SelectedVariables = [featureVariables; ens.ConditionVariables];

reset(ens)

idxResponse = strcmp(featureData.Properties.VariableNames,'SensorDrift');

idxLastFeature = find(idxResponse)-1; % Index of last feature to use as a potential predictor

featureAnalysis =    fscnca(featureData{:,1:idxLastFeature},feature Data.SensorDrift);

featureAnalysis.FeatureWeights

idxSelectedFeature = featureAnalysis.FeatureWeights > 0.1;

classifySD = [featureData(:,idxSelectedFeature), featureData(:,idxResponse)]

figure

histogram(classifySD.SigRangeCumSum(classifySD.SensorDrift),'BinWidth',5e3 )

xlabel('Signal cumulative sum range')

ylabel('Count')

hold on

histogram(classifySD.SigRangeCumSum(~classifySD.SensorDrift),'BinWidth',5e 3)

hold off

legend('Sensor drift fault','No sensor drift fault')
```

7.1.2. Computer Vision System

Machine vision detection system uses CCD camera to convert the detected object into image signal, which is transmitted to a special image processing system. According to the information of pixel distribution, brightness, color and so on, it is transformed into digital signal. The image processing system performs various operations on these signals to extract the features of the target, such as area and quantity. According to the preset permissibility and other conditions, output the results, including size, angle, number, qualified/unqualified, yes/no, *etc.*, to realize the automatic identification function [2].

(1) Main algorithms

The machine vision detection system often uses image registration, object detection and classification, tracking, and motion estimation methods. The algorithms should do scale changes, picture rotation and occlusion. The main algorithms in Matlab toolkit include FAST [3], corner detectors [4], and the SURF [5]and MSER blob detectors [6]. The descriptors of the picture include the SURF, FREAK, BRISK, LBP, and HOG descriptors.

The other algorithms are introduced in the following:

a. K-Means: Algorithms for computing data aggregation, mainly by constantly taking the nearest mean of seed points.
b. Mean-shift method: generally, refers to an iterative step, that is, first calculating the migration mean of the current point, moving the point to its migration mean, and then taking it as a new starting point, continuing to move until certain conditions are satisfied.
c. Covariance Matrix method: Covariance reflects the correlation between two random variables.
d. SIFT: (Scale-invariant transform) is an algorithm for detecting local features. The algorithm is proposed by D.G. Lowe in 1999 and perfected in 2004. Later, Y. He improved the descriptor by replacing histogram with PCA. The SIFT algorithm is an algorithm for extracting local features. It finds extreme points in scale space, extracts location, scale and rotation invariant SIFT features. It keeps invariance to rotation, scale scaling, brightness change, and stability to a certain extent to view angle change, affine transformation and noise.
e. Regression problem method: Given multiple independent variables, a dependent variable and some training samples representing the relationship between them, how to determine their relationship? Modeling: The purpose is to find the function of this dependent variable with respect to these independent variables. And this function can more accurately express the relationship between the dependent variable and the multiple independent variables.

The processing of machine vision detection system using MATLAB is described in the following:

A. Feature Detection and Feature Extraction

The normal methods include SIFT, SURF, BRISK or FREAK descriptor.

B. Choose a Feature Detector and Descriptor

For example, the computer vision system toolbox using in Position Control and Product Detection of Gravure Printer:

Video images of printed productions are continuously captured by a camera set on the production line. The speed of the camera is below 30 frames/s and can be adjusted. First, the image captured by the camera is quantized, and the analog signal is converted into digital signal, from which a keyframe which effectively represents the content of the lens is extracted and displayed on the display. The analysis method of image can be used to process a frame of image. Through size measurement and multi-spectral analysis, the color markers on the video image can be identified, and the distance between the color markers and the color parameters of the color markers as well as some other correlations can be obtained.

Due to various factors, there will be a variety of noise, such as Gauss noise, pepper noise and random noise. The noise brings many difficulties to image processing. It has a direct impact on image segmentation, feature extraction and image recognition. Therefore, real-time collected images need to be filtered. Image filtering requires the removal of noise outside the image while maintaining the details of the image. When the noise is Gauss noise, the most commonly used is the linear filter, which is easy to analyze and implement; but the linear filter has a poor filtering effect on salt and pepper noise. The traditional median filter can reduce salt and pepper noise in the image, but the effect is not ideal, that is, the fully dispersed noise is removed, and the noise close to each other will be removed. So, when the salt and pepper noise is serious, its filtering effect is obviously worse. The improved median filtering method is used in this system. The method first obtains the median of the maximum and minimum gray value pixels removed from the noise image window, then calculates the difference between the median and the corresponding gray value of the pixel, and then compares it with the threshold to determine whether to replace the gray value of the pixel with the calculated value or not.

In this stage, image segmentation detects each color mark and separates it from the background. There are two kinds of edges reflected by gray discontinuity. One is a step edge. The gray values of the pixels on both sides are significantly different. The other is the roof edge, which is located in the change of gray values from increase to decrease. For step edges, the second-order directional derivative of breakpoint L is zero-crossed at the edge, so differential operator can be used as edge detection operator. Differential operator-like edge detection method is similar to high-pass filtering in the high spatial domain, which can increase high-frequency components. This kind of operator is very sensitive to noise. For step edges, generally available operators are gradient operator Sobel operator and Kirsh operator. Laplace transform and Kirsh operator can be used for roof edges. Because the color scale is rectangular and the gray levels of adjacent edges are quite different, edge detection is used to segment the image. Sorbet edge unit is used to detect edges. It uses local difference operator to find edges and can separate color labels better. In the actual detection process, the color image edge detection method is used, and the appropriate color base (such as intensity, chroma, saturation, *etc.*) is selected to detect. According to the typical characteristics of the printing press, *i.e.* the color and layout characteristics of the printing press, multi-threshold processing is carried out to obtain binary maps of different colors.

The segmented image is measured and the object is identified by the measured value. Because the color scale is a rectangle with regular shape, the following features can be extracted: (1) Rectangular area is calculated from the pixels, (2) Rectangularity, (3) Chromaticity (H) and Saturation (S). Then the number of pixels between the color scales is obtained according to the number of pixel points between the color scales. The difference between the two values is obtained by comparing the distance with the set value. The difference is measured m times, and the average difference is taken to provide the corresponding adjustment signal for the digital AC servo regulation part. To adjust the relative position of color rollers, thereby eliminating or reducing printing misalignment. In feature extraction, multi-spectral image analysis can quantitatively represent color markers, such as the color of pixels in color digital images. Two parameters of color information of each color marker, chroma, and saturation, can be obtained by HIS format to detect the quality of ink. Statistical calculation of each color binary map or template matching with standard graphics were carried out to measure the parameters such as ink chips in the printing process.

The printing press is unwinding by the uncoiler, which passes through each printing unit in turn, printing, and drying of various colors. Each color printing by the uncoiler will be printed on the edge of the printing material for color registration. The color marking line is 10 mm horizontal and 1 mm wide. When each adjacent color marking line is accurate, it should be parallel and vertical to each other. (Longitudinal) phase is huge 20 mm. Video images of printed products are continuously captured by cameras installed on the production line. Through size measurement and multi-spectral analysis, the color markers on the video images can be identified. The color parameters L of the color marker spacing and color markers are obtained. If the distance between adjacent two-color markers is greater than or less than 20 mm, the overprint deviation occurs. The deviation signal is sent to the servo frequency conversion drive unit to drive the AC servo motor so that the corresponding color correction roll ML moves up and down to extend or shorten the printing material's travel from the last printing roll to the printing roll of the unit to dynamically correct.

Example 7-2: Track pedestrians using a camera mounted in a moving car.

Solution 7-2:

% Create system objects used for reading video, loading prerequisite data file, detecting pedestrians, and displaying the results.

video File = 'vippedtracking.mp4';

scale Data File = 'pedScaleTable.mat'; % An auxiliary file that helps to determine the size of a pedestrian at different pixel locations.

obj = setupSystemObjects(videoFile, scaleDataFile);

detector = peopleDetectorACF('caltech');

% Create an empty array of tracks.

tracks = initializeTracks();

% ID of the next track.

nextId = 1;

% Set the global parameters.

option.ROI = [40 95 400 140]; % A rectangle [x, y, w, h] that limits the processing area to ground locations.

option.scThresh = 0.3; % A threshold to control the tolerance of error in estimating the scale of a detected pedestrian.

option.gating Thresh = 0.9; % A threshold to reject a candidate match between a detection and a track.

option.gating Cost = 100; % A large value for the assignment cost matrix that enforces the rejection of a candidate match.

option.costOfNonAssignment = 10; % A tuning parameter to control the likelihood of creation of a new track.

option.time Window Size = 16; % A tuning parameter to specify the number of frames required to stabilize the confidence score of a track.

option.confidence Thresh = 2; % A threshold to determine if a track is true positive or false alarm.

option.age Thresh = 8; % A threshold to determine the minimum length required for a track being true positive.

option.visThresh = 0.6; % A threshold to determine the minimum visibility value for a track being true positive.

% Detect people and track them across video frames.

stop Frame = 1629; % stop on an interesting frame with several pedestrians

for fNum = 1:stopFrame

frame = readFrame();

[centroids, bboxes, scores] = detectPeople();

predictNewLocationsOfTracks();

[assignments, unassignedTracks, unassignedDetections] = ...

detectionToTrackAssignment();

updateAssignedTracks();

updateUnassignedTracks();

deleteLostTracks();

createNewTracks();

displayTrackingResults();

% Exit the loop if the video player figure is closed.

if ~isOpen(obj.videoPlayer)

break;

end

end

7.1.3. Robotics System Toolbox

The main algorithms of Robotics System Toolbox™ include map representation, path planning, and path following for differential drive robots [7].

It is a Robot Software Platform that can provide work similar to operating systems for heterogeneous computer clusters.

The predecessor of ROS is the switchyard project set up by Stanford Artificial Intelligence Laboratory, it is supported by Stanford intelligent robot STAIR. ROS provides some standard operating system services, such as hardware abstraction, underlying device control, implementation of common functions, inter-process message and packet management. ROS is based on a graphical architecture, so that processes of different nodes can accept, publish and aggregate various information (such as sensing, control, state, planning, *etc.*). At present, ROS mainly supports Ubuntu.

ROS can be divided into two layers. The lower layer is the operating system layer described above, while the higher layer is the various software packages contributed by the vast user groups to achieve different functions, such as location mapping, action planning, perception, simulation and so on.

ROS (low-level) uses BSD licenses, all open source software, and can be used for research and commercial purposes free of charge. High-level user-provided packages can use many different licenses.

1) Introduction

Robot Operating System (ROS) is a communication interface that enables different parts of a robot system to discover, send, and receive data.

2) ROS Terminology

A *ROS network* includes a ROS master and ROS node. A *ROS master* coordinates the other components in the network. A *ROS node* is the primary element. It uses messages to exchange data, use *Publishers*, *subscribers*, and *services* methods. A publisher sends messages to a specific *topic*, and subscribers to that topic receive those messages. The process of Message communication is (Fig. **7-4**).

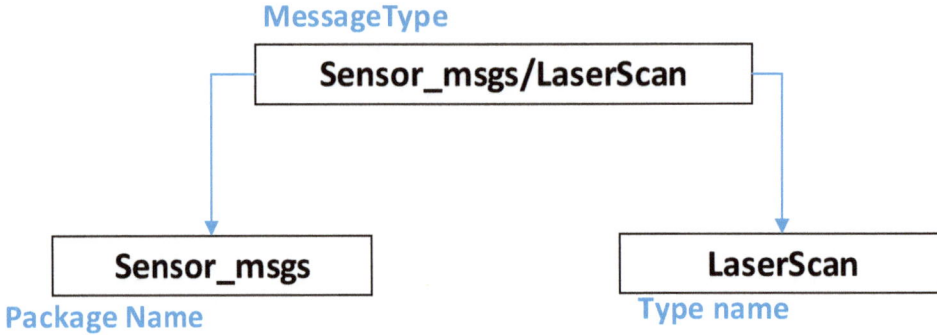

Fig. (7-4). The message type of ROS messages.

The primary mechanism for ROS nodes to exchange data is shown in Fig. (**7-5**).

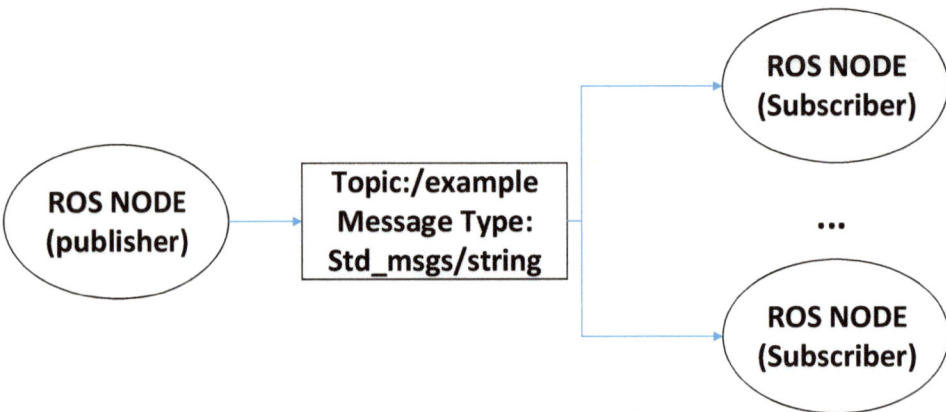

Fig. (7-5). The concept of topics, publishers, and subscribers.

Example 7-3: Exchange Data with ROS Publishers and Subscribers

Solution 7-3:

rosinit

exampleHelperROSCreateSampleNetwork

rostopic list

rostopic info /scan

```
laser = rossubscriber('/scan')

scandata = receive(laser,10)

figure

plot(scandata,'MaximumRange',7)

robotpose = rossubscriber('/pose',@exampleHelperROSPoseCallback)

global pos

global orient

pause(2)

pos

orient

clear robotpose

chatterpub = rospublisher('/chatter', 'std_msgs/String')

pause(2)

chattermsg = rosmessage(chatterpub);

chattermsg.Data = 'hello world'

rostopic list

chattersub = rossubscriber('/chatter', @exampleHelperROSChatterCallback)

send(chatterpub,chattermsg)

pause(2)

exampleHelperROSShutDownSampleNetwork

clear global pos orient

rosshutdown

displayEndOfDemoMessage(mfilename)
```

7.2. ARDUINO IN INTELLIGENT INSTRUMENT

Arduino is a convenient, flexible and user-friendly open source electronic prototype platform. Includes hardware (various models of Arduino board) and software (Arduino IDE). Developed by a European development team in the winter of 2005. Its members include Massimo Banzi, David Cuartielles, Tom Igoe, Gianluca Martino, David Mellis and Nicholas Zambetti.

It is built on the open source simple I/O interface and has a Processing/Wiring development environment similar to Java and C. There are two main parts: the hardware part is the Arduino circuit board which can be used as circuit connection; the other part is the Arduino IDE, the program development environment in your computer. You just write the program code in the IDE and upload it to the Arduino circuit board. The program will tell the Arduino circuit board what to do.

Arduino can sense the environment through a variety of sensors, and feedback and influence the environment by controlling lights, motors and other devices. The microcontroller on the board can be programmed by Arduino programming language, compiled into binary files, and burned into the microcontroller. Arduino programming is achieved through Arduino programming language (based on Wiring) and Arduino development environment (based on Processing).

7.2.1. SLAM

SLAM problem can be described as: the robot moves from an unknown position in an unknown environment, locates itself according to position estimation and map in the process of moving, and builds an incremental map on the basis of self-localization to realize the autonomous localization and navigation of the robot.

Intelligent SLAM means Semantic SLAM, Accurate Sensing and Adaptation to Environment. Semantic analysis and SLAM are effectively integrated to enhance the understanding ability of machine interaction in the environment, which endows the robot with complex environment perception and dynamic scene adaptability. SLAM with breadth has 1 million square meters strong mapping capability, With the help of efficient environment recognition and intelligent analysis technology, the robot will have the ability to construct maps with a full range of indoor and outdoor scenes up to 1 million square meters.

Precision SLAM: Leading High Precision Location Algorithms, SLAM2.0 can recognize and locate the power-on at any location with a precision of 2 cm.

Time-efficient SLAM: Real-time updating of dynamic maps, according to the data returned by the sensor, the analysis and comparison with the original map are carried out to complete the dynamic real-time update and realize life-long SLAM [9].

Mathematically, the map is represented by a state vector $\hat{x}$ and covariance matrix **P**. State vector $\hat{x}$ is composed of the stacked state estimates of the camera and features, and **P** is a square matrix of equal dimension which can be partitioned into sub-matrix elements as follows:

$$
\hat{x} = \begin{pmatrix} \widehat{x_v} \\ \hat{y}_1 \\ \hat{y}_2 \\ ... \end{pmatrix}, \mathbf{P} = \begin{bmatrix} P_{xx} & P_{xy1} & P_{xy2} & & \\ P_{y1x} & P_{y1y1} & P_{y1y2} & .. & .. \\ P_{y2x} & P_{y2y1} & P_{y2y2} & & \\ & & .. & & \\ & & & .. & \end{bmatrix} \tag{1}
$$

Explicitly, the camera's state vector x_v comprises a metric 3D position vector r^w, orientation quaternion q^{Rw}, velocity vector v^w and angular velocity vector ω^R relative to a fixed world frame W and 'robot' frame R carried by the camera:

$$
\mathbf{X}_v = \begin{pmatrix} r^W \\ q^{WR} \\ v^{\ W} \\ \omega^R \end{pmatrix} \tag{2}
$$

The role of the map is primarily to permit real-time localization rather than serve as a complete scene description and therefore aim to capture a sparse set of high-quality landmarks.

Kinect: As we all know; this is a depth camera. You may have heard of other brands, but Kinect is cheap, measuring range is between 3m and 12m, accuracy is about 3cm, more suitable for indoor robots. It takes images like this (RGB, depth and point clouds from left to right).

One of the advantages of Kinect is that it can get the depth value of each pixel cheaply, both in terms of time and economy. OK, with this information, the radish actually knows the 3D position of each point in the picture it captures. As long as we calibrate Kinect beforehand or use factory calibration values.

The 's' on the left is the scale factor, which means that all the rays coming out of the camera's light center will fall at the same point on the imaging plane. If we don't know the distance, then 's' is a free variable. But in RGB-D cameras, we know this distance in the Depth diagram, and its reading dep (u, v) is a multiple of the true distance. If also recorded as 's'.

The method of calculating three-dimensional points: First read the depth data from the depth map (Kinect gives 16-bit unsigned integers) and remove the zoom factor in the Z direction, so that you can change an integer into data in meters. Then X and Y are calculated using the above formula. It's not difficult at all. It's just a central point and a focal length. F stands for focal length and C stands for the center. If you don't calibrate your Kinect yourself, you can also use the default values: s = 5000, CX = 320, CY = 240, FX = FY = 525. The actual value will be a little biased, but not too large.

Knowing the location of each point in Kinect, what we need to do next is to calculate the displacement of radish according to the difference between the two images. For example, the following two pictures, the later one was collected one second after the previous one.

For relative attitude estimation of the camera, the classical algorithm is ICP (Iterative Closest Point). This algorithm requires to know a set of matching points between the two images. The popular point is which points in the left image are the same as those in the right image. Of course, you can see that black and whiteboard appears in both images at the same time. In the view of radish, there are two simple problems involved here: feature extraction and matching.

The algorithms include SIFT, SURF and other features. Yes, to solve the problem of location, we first need to get a match between the two images. The basis of matching is the feature of the image.

After we get a set of matching points, we can calculate the transformation relationship between two images, also known as PnP problem.

The motion equation describes how the robot moves. U is the input of the robot, W is the noise. The simplest form of this equation is how you can get the displacement difference between two frames (code disk, *etc.*). Then the equation is directly obtained by adding the last frame to u. In addition, you can also completely do without inertial measurement equipment, so we only rely on image equipment to estimate, which is also possible.

The latter equation, called the observation equation, describes how those landmarks came about. You see the jth landmark in frame I and produce a measurement value, the abscissa and vertical coordinates in the image. The last item is the noise. Kalman filter still occupies the most important position in SLAM system. Davison's classical SLAM is made with EKF.

Example 7-4: SLAM and autonomous navigation with ROS + Kinect + Arduino + android [10]

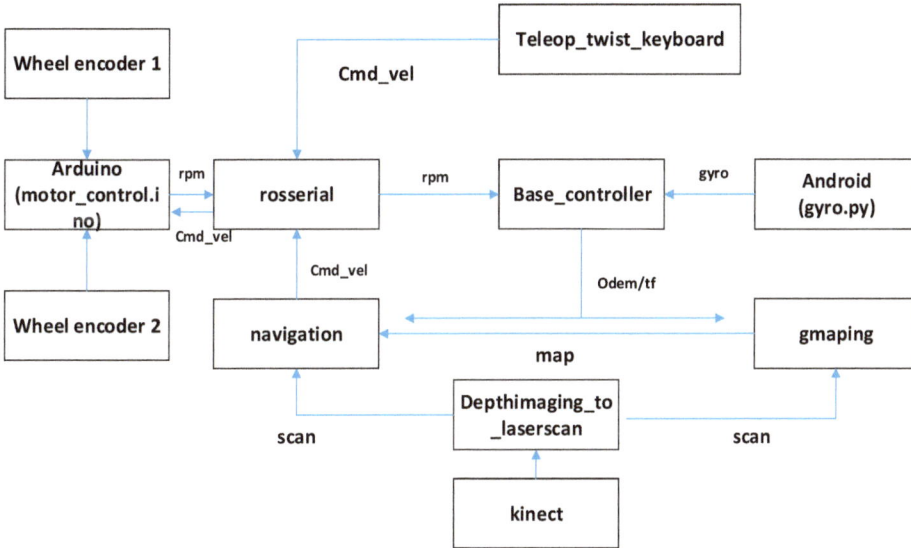

Fig. (7-6). The flowchart of slam based on Arduino platform [10].

Solution 7-4:

1) The Hardware:

2) Arduino(motor_controller.ino):

The main loop in the motor controller. It converts encoder tick counts into the current rpm of both motors and controls the speed of the motors using PID control.

For example, the C code in Adafruit_MotorShield.cpp

void Adafruit_DCMotor::setSpeed(uint16_t speed)

{

MC->setPWM(PWMpin, speed);

}

The uint8_t parameter in the corresponding declaration in Adafruit_MotorShield.h also needs to be modified accordingly. The Adafruit motor shield v2 has its own on-board PWM chip with 12-bit resolution, but for some reason the dc motor library only used 8 bits.

3)base_controller.cpp

The base_controller subscribes to the rpm topic from the Arduino and converts it into x velocity, theta velocity, xy position, and yaw.

4)Android(gyro.py):

It is necessary to add an IMU to the robot to improve odometry

6) Kinect:one can follow the instructions in

http://wiki.ros.org/kinect/Tutorials/Adding%20a%20Kinect%20to%20an%20iRo bot%20Create

5) teleop_twist_keyboard:

The teleop_twist_keyboard ros package takes in keyboard input and publishes cmd_vel messages.

6) G mapping:

7.2.2. PHM

The dimensionality reduction techniques, with some of the key concepts associated with data analysis and visualization, for applications of both linear and non-linear data sets, also for time sensitive networks and storage sensitive networks (*e.g.*, IoT, edge computing, big data). It is an important method of PHM [11].

Power drives are widely used in many fields, for example, solar inverters, wind farms, grid-tied battery system, electric modern trains and electrical vehicles. The increasing penetration of power drives makes their reliability, robustness, and early diagnosis a central point of attention especially in planning, designing, and financing.

Its fault diagnosis includes the following topics:

- Life expectancy and failures mode as a holistic approach in component selection;
- Material degradation in semiconductors, conductor and dielectrics;
- Capacitors;
- Batteries;
- Inverter topologies;
- Early diagnosis;
- Fault-tolerant software strategies.

In classical reliability engineering and life data analysis, the state of an item is binary: survive or fail. The number of each state, or the number of failures or survivors, is always an integer. For example, a 6-week life testing of 100 units yielded 5 failures and 95 survivors.

However, in real world practice, fractional failures, and therefore fractional survivors as well, can be encountered in various scenarios. For example, the number of chargeable failures is fractional when:

- Corrective actions (CA) are partially effective (less than 100%),
- Performance degradation has crossed the pre-specified threshold but hasn't manifested as a physical failure yet,
- Initial failure analysis cannot duplicate the field failure symptom due to failure diagnosis limitation,
- the actual root cause is not readily available for each failure due to lack of FA resource, but there is empirical knowledge for the likelihood for each potential root cause, *etc.*
- The consequence (harm) of a failure or hazard is diversified and uncertain, but the risk needs to be assessed for each, *etc.*

It may be seen the above scenarios share one thing in common; *i.e.*, the state of the subject in the study is not known or uncertain. It is normal cases in risk assessment, and decision making, to properly classify, model, and to handle the data using reliability analysis software. For example, the MLE parameter estimation can already handle such scenarios (but with different data entry formats). The reliability practitioners realize the existence of the fractional failures and survivors, quantify their likelihood, enter the data correctly, and interpret the results properly.

The commercial software, such as Weibull++, JMP, and Minitab, provides numerical examples to demonstrate the actual application in real world practice including data collection, failure classification, and fractional failure determination, data entry format, life distribution parameter estimation, reliability quantification, and field risk prediction.

Meanwhile, over the past several years, a paradigm shift has occurred in engineering fields where complex systems are presenting more characters of hybrid, interaction, dynamic, data-rich and multi-energy transformation. Prognostics and Health Management (PHM) is essential in guaranteeing the safe, efficient, and correct operation of the complex of detection, isolation, and identification of faults; and prognosis, which consists of prediction of the remaining useful life (RUL) of components, subsystems, or systems; constitute systems health monitoring.

PHM aims to provide users with an integrated view of the health state of equipment or the overall system. An effective PHM system is expected to provide early and isolation of the precursor and/or incipient fault of detection components or sub-elements; to have the means to monitor and predict the progression of the fault, and to aid in making, or autonomously trigger maintenance schedule and asset management decisions or actions.

The sensors used by PHM in Power drives normally are camera, infrared array sensor, UV sensor, and microphone. The infrared camera is a simple way to realize it.

Example 7-5: Infrared Array Sensor

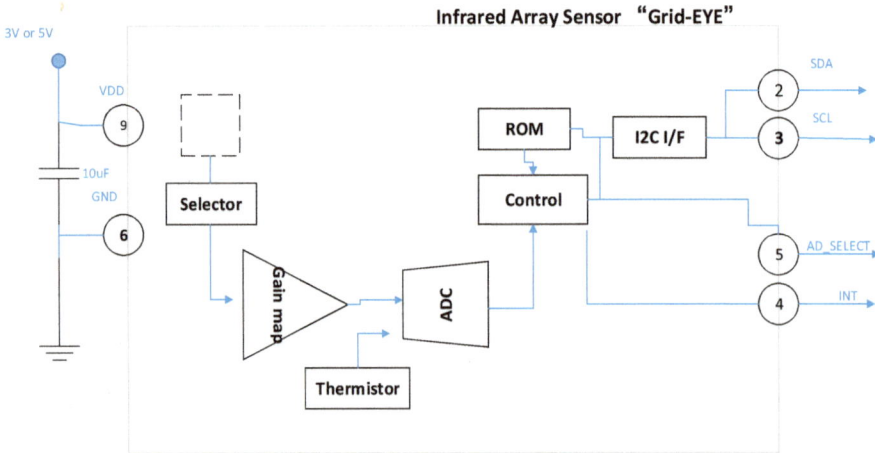

Fig. (7-7). The structure diagram of Infrared Array Sensor AMG8833 [12].

Solution 7-5:

This sensor from Panasonic is an 8x8 array of IR thermal sensors. When connected to your microcontroller (or raspberry Pi) it will return an array of 64 individual infrared temperature readings over I2C. It's like those fancy thermal cameras, but compact and simple enough for easy integration.

This part will measure temperatures ranging from **0°C to 80°C (32°F to 176°F)** with an accuracy of +- 2.5°C (4.5°F). It can detect a human from up to 7 meters (23) feet. With a maximum frame rate of 10Hz, it's perfect for creating your own human detector or mini thermal camera.

You can easily wire this breakout to any microcontroller; we'll be using an Arduino. You can use any other kind of microcontroller as well as long as it has I2C clock and I2C data lines.

Connect Vin to the power supply, 3-5V is fine. Use the same voltage that the microcontroller logic is based on. For most Arduinos, that is 5V

Connect GND to common power/data ground

Connect the SCL pin to the I2C clock SCL pin on your Arduino. On an UNO & '328 based Arduino, this is also known as A5, on a Mega it is also known as digital 21 and on a Leonardo/Micro, digital 3

Connect the SDA pin to the I2C data SDA pin on your Arduino. On a UNO & '328 based Arduino, this is also known as A4, on a Mega it is also known as digital 20 and on a Leonardo/Micro, digital 2

By default, the I2C address is 0x69. If you solder the jumper on the back of the board labeled "Addr", the address will change to 0x68.

Example python code for Adafruit_AMG88xx import Adafruit_AMG88xx.

import pygame

import os

import math

import time

import NumPy as np

from scipy.interpolate import griddata

from color import Color

```
#low range of the sensor (this will be blue on the screen)

MINTEMP = 26

#high range of the sensor (this will be red on the screen)

MAX TEMP = 32

#how many color values we can have

COLOR DEPTH = 1024

os.putenv('SDL_FBDEV', '/dev/fb1')

pygame.init()

#initialize the sensor

sensor = Adafruit_AMG88xx()

points = [(math.floor(ix / 8), (ix % 8)) for ix in range(0, 64)]

grid_x, grid_y = np.mgrid[0:7:32j, 0:7:32j]

#sensor is an 8x8 grid so lets do a square

height = 240

width = 240

#the list of colors we can choose from

blue = Color("indigo")

colors = list(blue.range_to(Color("red"), COLORDEPTH))

#create the array of colors

colors = [(int(c.red * 255), int(c.green * 255), int(c.blue * 255)) for c in colors]

displayPixelWidth = width / 30

displayPixelHeight = height / 30

lcd = pygame.display.set_mode((width, height))

lcd.fill((255,0,0))
```

```
pygame.display.update()

pygame.mouse.set_visible(False)

lcd.fill((0,0,0))

pygame.display.update()

#some utility functions

def constrain(val, min_val, max_val):

return min(max_val, max(min_val, val))

def map(x, in_min, in_max, out_min, out_max):

return (x - in_min) * (out_max - out_min) / (in_max - in_min) + out_min

#let the sensor initialize

time.sleep(.1)

while(1):

#read the pixels

pixels = sensor.readPixels()

pixels = [map(p, MINTEMP, MAXTEMP, 0, COLORDEPTH - 1) for p in pixels]

#perdorm interpolation

bicubic = griddata(points, pixels, (grid_x, grid_y), method='cubic')

#draw everything

for ix, row in enumerate(bicubic):

for jx, pixel in enumerate(row):

pygame.draw.rect(lcd, colors[constrain(int(pixel), 0, COLORDEPTH- 1)],
(displayPixelHeight * ix, displayPixelWidth * jx, displayPixelHeight,
displayPixelWidth))

pygame.display.update()
```

PROBLEM

7-1 design an intelligent instrument, its arithmetic using MATLAB, hardware using ARDUINO.

REFERENCES

[1]   R.K. Mobley, "6–Predictive Maintenance Techniques", *An Introduction to Predictive Maintenance,* vol. 8, pp. 99-113, 2002.
[http://dx.doi.org/10.1016/B978-075067531-4/50006-3]

[2]   P.I. Corke, "The Machine Vision Toolbox: a MATLAB toolbox for vision and vision-based control", *IEEE Robot. Autom. Mag.,* vol. 12, no. 4, pp. 16-25, 2005.
[http://dx.doi.org/10.1109/MRA.2005.1577021]

[3]   B. Krishnapuram, L. Carin, M.A. Figueiredo, and A.J. Hartemink, "Sparse multinomial logistic regression: fast algorithms and generalization bounds", *IEEE Trans. Pattern Anal. Mach. Intell.,* vol. 27, no. 6, pp. 957-968, 2005.
[http://dx.doi.org/10.1109/TPAMI.2005.127] [PMID: 15943426]

[4]   J. Shi, and C. Tomasi, "Good Features to Track", *IEEE Conference on Computer Vision & Pattern Recognition,* 2002

[5]   Marek Kraft, and A. Schmidt, "Simplifying SURF Feature Descriptor to Achieve Real-Time Performance", *Computer Recognition Systems 4,* 2011.
[http://dx.doi.org/10.1007/978-3-642-20320-6_45]

[6]   K.P. Jr, "White, and S. Robinson. "The problem of the initial transient (again), or why MSER works", *J. Simul.,* vol. 4, no. 4, pp. 268-272, 2010.
[http://dx.doi.org/10.1057/jos.2010.19]

[7]   P. Corke, "Robot manipulator capability in MATLAB: A Tutorial on Using the Robotics System Toolbox [Tutorial]", *IEEE Robotics & Automation Magazine PP.,* vol. 99, pp. 1-1, 2017.
[http://dx.doi.org/10.1109/MRA.2017.2718418]

[8]   Arduino programming language,, http://www.arduino.cc

[9]   A.J. Davison, I.D. Reid, N.D. Molton, and O. Stasse, "MonoSLAM: real-time single camera SLAM", *IEEE Trans. Pattern Anal. Mach. Intell.,* vol. 29, no. 6, pp. 1052-1067, 2007.
[http://dx.doi.org/10.1109/TPAMI.2007.1049] [PMID: 17431302]

[10]   SLAM and autonomous navigation with ROS + Kinect + Arduino + android,, Available: https://blog.csdn.net/linuxarmsummary/article/details/52527992,access

[11]   P.P. Bonissone, and N. Iyer, *Soft Computing Applications to Prognostics and Health Management (PHM): Leveraging Field Data and Domain Knowledge. Computational and Ambient Intelligence.* 2007. *AMG8833 datasheet.* Panasonic Corporation Automation Controls Business Unit, 2011.

## CHAPTER 8

The Application of Intelligent Instrument: A Mobile Intelligent Instrument

Abstract: In this chapter, the advances of the wearable and IoT intelligent instrument are discussed; some state-of-the-art instruments and their applications will be also introduced.

Keywords: ADS1692R, Body area network, CC2650, IoT, Vector network analyzer, Wearable intelligent instrument.

8.1. THE WEARABLE INTELLIGENT INSTRUMENT

The wearable intelligent instrument allows real-time monitoring of the health status of human beings, it is considered as an important approach to build health management and monitoring platforms. With the increasing technology of smart textiles, flexibility devices, FPCB and on-body antenna, the platform became more intelligent and interoperable [1].

The main issues that had been discussed in the publications include the IoT platforms, the wearable device's technology, the wireless body area network (WBAN) regulations, the channel models for WBAN in wireless medical communications, and so on [2].

To design a BAN smart instrument, it is better to follow the WBAN regulations, for example, 1) the transmissions distance is about 2 m; 2) density is about 2-4 nodes in 1m2; 3) Latency (end to end) is about 10ms, and so on.

Nowadays, test technology in the smart vest includes 1) Heart rate (HR). It is often be extracted from the ECG or PPG signals [3]. 2) Respiration rate (RR), the main methods are elastomeric plethysmography (EP), impedance plethysmography (IP) and respiratory inductive plethysmography. 3) Blood Oxygen Saturation, it is normal to measure using photoplethysmography (PPG) technology and pulse oximetry principles. 4) skin sweat .it is often tested by epidermal-based sensors and Fabric/flexible plastic-based sensors. 5) Body Temperature includes core temperature (CT) and skin temperature. 6) ECG, it often uses wet or dry electrodes, and textile-based technology, and so on.

8.1.1. The Body Temperature Sensors

There are two kinds of body temperature sensors: uncontacted or contact sensor. They are all high precision temperature RTD, IR, NTC/PTC thermometer sensors that have specially been designed for body temperature test. The LMT70, MAX30205, MLX90615 are three current devices often selected for body temperature. the body temperature test data from LMT70 and MAX30205 has relatively high accuracy when it contacts the skin for a certain time, and IR temperature sensors MLX90615 is easy to use but need to process its data carefully.

1) LMT70 [4]

LMT70 is comprised of stacked BJT base emitter junctions (sensing part), current source, a precision amplifier and the output switch.

To obtain the trusted data or process the test data, four influence factors should be considered: the self-heating; different power supply voltage; output resistance and load circuit; thermal response time.

A. The self-heating

In LMT70, its junction temperature is the actual measurement temperature. But the power dissipation of the device can induce the rise of a device junction temperature, it is calculated by Equation 8-1. The parameter is the thermal resistance junction to ambient ($R_{\theta JA}$).

$$T_J = T_A + R_{\theta JA}[(V_{DD}I_Q) + (V_{DD} - V_{TEMP})I_L] \tag{8-1}$$

Where

• T_A is the ambient temperature.

• I_Q is the quiescent current.

• I_L is the load current on VTEMP.

• I_J is the junction temperature.

• V_{DD} is the voltage of power.

• I_{DD} is the operating current of the device.

• V_{Temp} is the output voltage.

B. Different power supply voltage:

The different power supply voltage, the TA0 output different voltage, it is calculated by the formula (8-2).

$$T_A = T_{A\_2.7v} + (V_{DD} - 2.7) * Kt_T]$$ **(8-2)**

Where

• T_A is the temperature output voltage in TAO under VDD (power supply voltage).

• $T_{A\_2.7}$ is the temperature output voltage in TAO under 2.7V.

• Kt_T is the V_{APSS} sensitivity. It equals 2 typically, equal -9 minimum, and 8 maximums.

C: load circuit design.

In formula (8-1), to decrease the rise of a device junction temperature coming from self-heating, it should decrease its load current.

R_{OUT} Output Resistance (28TYP 80MUX)

T_{AO} Off Leakage Current($\pm$0.5uA)

Step 4: Thermal response time.

Thermal response time to 63% of final value in stirred oil 1.5sec.

Thermal response time to 63% of final value in still air 73sec. The thermal response time of the body temperature test is between 1.5sec to 73sec as its initial status.

2) MAX30205 [5]

The MAX30205 output temperature data directly, and it is easy to do multipoint test though IIC bus, (the LMT70 use multi AD convert port). Its data features and principle are similar as LMT70.

3) MLX90615 Infra-Red Thermometer

The MLX90615 IR sensor includes thermo-couples, thick chip substrate (cool terminal), thin membrane (hot terminal). The thermopile output signal is calculated using the formula (8-3).

$$V_{ir}(T_A, T_O) = A \cdot (T_0 - T_A) \tag{8-3}$$

Where T_O represents the object absolute temperature (Kelvin), TA represents the sensor die absolute (Kelvin) temperature, and A is the overall sensitivity.

Its data has the following characteristics:

As a wearable device in a vest, its test distance is varied slightly when the body is moved, this produces a slightly changed data. In a small range the accuracy can be improved by correction.

8.1.2. The Propitiation Sensors

In the propitiation test, the propitiation chemical biosensor provides many valuable test data when it contacts to the sweat, but it is normal, not easy in a smart vest. In this paper, we present the method that used humidity sensor SHT20 to test propitiation.

1) SHT20 [6]

The SHT20 contains a capacitive type humidity sensor, it calculates relative humidity use formula (8-4) and (8-5).

$$RH = -6 + 125 \cdot \frac{S_{RH}}{2^{16}} \tag{8-4}$$

The relative humidity signal output is SRH, the relative humidity is RH.

$$RH_i = RH_w \cdot \frac{\exp(\frac{\beta_w \cdot t}{\lambda_w + t})}{\exp(\frac{\beta_i \cdot t}{\lambda_i + t})} \tag{8-5}$$

For relative humidity above ice RHi the values need to be transformed from relative humidity above water RHw at temperature t. The corresponding coefficients are defined as follows: $\beta w = 17.62$, $\lambda w = 243.12°C$, $\beta i = 22.46$, $\lambda i = 272.62°C$.

In the vest application, its response time is nearly 1-10 seconds, typically 2 -4 seconds.

2) Potential chemical biosensor and camera

The potential sensor uses sensor array, they are designed for situ perspiration analysis, for example, the sweat metabolites, electrolytes and the skin temperature. The camera is also situ perspiration analysis sensors in the feature.

8.1.3. The SpO2 Test (PPG) Sensors

There are two kinds of Pulse Oximeters, one is transmitted light and photodetector on the different side of the finger; the other is transmitted light and photodetector on the same side. The front one is stable and is often used for clinical application, the behind one is relatively loose and often used for health monitoring. The MAX30102 is the behind one, the AFE4403 is used in the front one (in finger clip).

1) SpO2 test

An LED light of different wavelength emitted and traveled through tissue, venous blood, and arterial blood, then it is collected. The transmitted light changes with time for the heartbeat induced the flow of blood. Normally the red and infrared lights are used for pulse oximetry to estimate the true hemoglobin oxygen saturation of arterial blood. Oxyhemoglobin (HbO2) absorbs visible and infrared (IR) light differently than deoxyhemoglobin (Hb) and appears bright red as opposed to the darker brown Hb. Then the SpO2 is calculated.

Its SpO2 subsystem contains ambient light cancellation (ALC), a sigma-delta ADC, and a discrete time filter. Meanwhile, the MAX30102 has an on-chip temperature sensor for calibrating the temperature dependence of the SpO2 subsystem.

In the process of calculating SpO2, absorption in the arterial blood is represented by an AC signal which is superimposed on a DC signal representing absorptions in other substances.

To simplify design, the vest choice MAX30102 to test the SpO2 value, and use AFE4403 to verify its data.

2) Heart rate

The heart rate is processed from the test data from the MAX30102, it is obvious, the time interval of heart rate (or effective test interval) is longer than the SpO2.

The mobility changes the heart rate largely when it is loose.

But in a smart vest,, the MAX30102 is a better choice for its wear comfortable and easy use.

3) The MAX30101 of the vest uses PPG arithmetic [7].

Firstly, the Beer-Lambert Law (formula 8-6) is used to calculate the intensity of the transmitted light.

$$I = I_0 \cdot e^{-\varepsilon(\lambda).C.L} \tag{8-6}$$

Where:

I is the intensity of transmitted light;

I_0 is the intensity of incident light;

C is the concentration of absorbent, mol (mol);

L is optical path length in the cm.

ε is absorptivity (extinction coefficient) of the substance at a specific wavelength, mol^{-1}/cm^{-1} (1/mol centimeters)

$$A = -ln\frac{I}{I_0} = -\varepsilon(\lambda).C.L \tag{8-7}$$

Equation 11 shows the calculation of Absorbance, the formula (8-8) shows that Absorbance is an additive function Absorbance of a mixture is a sum of the absorbances of the components.

$$A = \varepsilon_X[X]L + \varepsilon_Y[Y]L \tag{8-8}$$

where [X] and [Y] are Deoxy-hemoglobin and Oxy-hemoglobin. So, absorption of the mixture at a wavelength $\lambda 1$ is shown in formula (8-9).

$$A_1 = \varepsilon_1{}^{Hb}[Hb]L += \varepsilon_1{}^{HbO_2}[HbO_2]L \tag{8-9}$$

The absorption of the mixture at a wavelength $\lambda 2$ is shown in formula (8-10).

$$A_2 = \varepsilon_2{}^{Hb}[Hb]L += \varepsilon_2{}^{HbO_2}[HbO_2]L \tag{8-10}$$

Secondly, use the following steps to calculate the SpO2

a) Measure the molar absorptivity of deoxyhemoglobin and oxyhemoglobin at two wavelengths;

b) Measure absorbance, the mixture at two wavelengths;

c) Solve the system;

$$HbO_2 = \frac{\varepsilon_1{}^{Hb} \cdot A_2 - \varepsilon_2{}^{Hb} \cdot A_1}{\varepsilon_2{}^{HBO_2} \cdot \varepsilon_1{}^{Hb} - \varepsilon_2{}^{Hb} \cdot \varepsilon_1{}^{HBO_2}} \qquad (8\text{-}11)$$

where [Hb02] is the concentration of oxy-hemoglobin.

$$Hb = \frac{\varepsilon_2{}^{HbO_2} \cdot A_1 - \varepsilon_1{}^{HbO_2} \cdot A_2}{\varepsilon_2{}^{HBO_2} \cdot \varepsilon_1{}^{Hb} - \varepsilon_2{}^{Hb} \cdot \varepsilon_1{}^{HBO_2}} \qquad (8\text{-}12)$$

where SpO2 is the oxygen saturation in the blood.

$$SpO_2 = \frac{HbO_2}{HbO_2 + Hb} \qquad (8\text{-}13)$$

Thirdly, use the experience formula (8-14) to verify (8-12)Measure absorbance, the mixture at two wavelengths.

$$R = \frac{\frac{I_{peak}(ac)}{I(dc)}(RED)}{\frac{I_{peak}(ac)}{I(dc)}(IR)} \qquad (8\text{-}14)$$

Where I_{peak} is the peak amplitude of intensity of light.

$$SpO2\% = 112.5 - R*25 \qquad (8\text{-}15)$$

8.1.4. The Respiration Rate Test Sensors

There are a few productions used for respiration rate test. The DLCK365 module provides trusted test data in a smart vest, the potential use includes the non-contact capacitive sensor and thermal imaging in the following part.

1) Abdominal respiration module DLCK365

Abdominal respiration module DLCK365 Fig. (**8-1**) is an abdominal breathing module assessment Kit. Its power supply range is 3.3V-5.0V, it can test

Respiration Rate 10-40 times per minute with the interval error $\leq\pm$ 3 times per minute. The pressure range is between 0-299mHg, error $\leq\pm$3mmHg. The resolution of pressure is 1mmHg.

Its principle is that it tests the pressure in blocking airbag, then calculate the abdominal respiration and strength. It is composed of an abdominal test module, Airbags (including hose), the fixation band, and hose components.

2) Non-contact capacitive sensing

The capacitive sensor technology has been used for the non-obstructive monitoring of the respiratory rate. It measures the capacitance existing between two metal plates (electrodes) together with the thoracic tissue acting as a dielectric material. The mechanical changes produced by breathing cause variations in the capacitance, it can be tested using on an LC oscillator (the letter L represents an inductor, and the letter C stands for a capacitor) to accurately measure the capacitance.

3) Thermal imaging

The use of thermal imaging for monitoring respiration is promising and could be more convenient for both patients and health practitioners. However, the use of the captured thermal images to monitor breathing is not straightforward and requires further processing before applying the measurement techniques.

So, to simplify design, the vest choice Abdominal respiration module DLCK365 to test the PPG test data.

8.1.5. The ECG Test

Electrocardiography (ECG), electromyography (EMG), and electroencephalography (EEG) systems measure heart, muscle, and brain activity (respectively) over time by measuring electric potentials on the surface of living tissue. The ECG data is provided by using test sensor AD8233, BMD101.

The BMD101 [8] is single chip solution for accurate Biosignal detection and processing for the ECG test. Its front circuit is similar to the AD8233.

To provide the valuable test data, the vest choice the BD101 chip to output the ECG waveform. And its results are used to verify the test data from AD8233.

Firstly, The range of low frequency is from 0.1Hz to 2Hz. The range of high frequency is larger than 50Hz. The main filter algorithm includes the low pass filter, the LMS adaptive filter, the ICA (independent component analysis) algorithm is introduced. the ICA algorithms use Kurtosis, Entropy. The Kurtosis is calculated to use formula (8-16). A weighted vector W in adaptive filter is estimated by kurtosis or entropy using formula (8-17). The renew coefficient of adaptive filer is using formula (8-18).

$$K(\text{kurtosis}) = \frac{\frac{1}{N}\sum_{t=1}^{N}(\bar{Y}-Y^t)^4}{\left(\frac{1}{N}\sum_{t=1}^{N}(\bar{Y}-Y^t)^2\right)^2} - 3 \tag{8-16}$$

$$W_{NEW} = W_{old} + \mu(4E[Z(W_{old}Z)^3]) \tag{8-17}$$

$$Y = WZ \tag{8-18}$$

• W_{NEW} is the new weighed vector, W_{OLD} is the last weighed vector before WNEW, W is weighted matrix.

• N is test number, Y is estimated signal, $\overline{Y}$ is its average value and the Y_t is its estimate value.

•µis renew coefficient, and Z is extracted unit length eigenvector, which is mutually orthogonal to X (the test signal).

The simulate database includes: MIT-BIH Arrhythmia Database; MIT-BIH Long-Term ECG Database; and Practical wearable ECG datasets use AD8832.

Secondly, the BD101 output its ECG waveform through UART port, this digital data includes its measurement error and data transmitted error. We use integrity checking the filter, limiting filter and peak to peak detection to process the ECG waveform.

8.2. THE IOT INTELLIGENT INSTRUMENT

Internet of Things (IoT) refers to real-time acquisition of any object or process that needs to be monitored, connected and interacted through various information sensors, radio frequency identification technology, global positioning system, infrared sensors, laser scanners and other devices and technologies, and acquisition of its sound, light, heat, electricity, mechanics, chemistry, biology, location and other needs. The necessary information can be accessed through various possible networks to realize the ubiquitous connection between things and people and realize the intelligent perception, recognition and management of goods and processes. The Internet of Things (IoT) is a carrier of information based on the Internet, traditional telecommunication network and so on. It enables all ordinary physical objects that can be independently addressed to form interconnected networks [9].

The typical architecture of the Internet of Things is divided into three layers: perception layer, network layer, and application layer. The core capability of the perception layer to realize the overall perception of the Internet of Things is the key technology, standardization, and industrialization of the Internet of Things. The key is to have more accurate and comprehensive perception ability, and to solve problems of low power consumption, miniaturization and low cost. The network layer is mainly based on the mobile communication network which is widely covered. It is the most standardized, industrialized and mature part of the Internet of Things. The key is to optimize the application characteristics of the Internet of Things and form a system-aware network. The application layer provides abundant applications, combines Internet of Things technology with industry information Imation needs and realizes a wide range of intelligent application solutions. The key lies in industry integration, development, and utilization of information resources, low-cost and high-quality solutions, the security of information and the development of effective business models [9].

8.2.1. The Power Instrument Used in the IoT

Many sensor nodes in IoT make the power measurement important. The main parameter is Power and Energy. It is shown in (8-19) and (8-20).

$$\text{Power} = I \times V \tag{8-19}$$

$$\text{Energy} = I \times V \times \text{Time} \tag{8-20}$$

Until recently, no single instrument offered the combination of low current measurement range, resolution, and speed capability necessary [10].

In the early stage of radio development, most of the test engineers are faced with the continuous wave, AM, FM, phase modulation or pulse signals, which can be followed regularly. For example, the power measurement of CW, FM or PM signals is simple, only the average power is needed to be measured; the power of AM signals is related to their modulation depth, and the characteristics of pulse signals are expressed in terms of pulse width and duty cycle. For these analog or analog modulated signals, the average power and peak power are the main concerns of RF power measurement.

The envelope of the digital modulated signal is irregular, and its maximum and minimum levels will change randomly, and the amount of change is large. In order to describe the characteristics of this kind of signal, some new descriptive methods are introduced, such as lead power, burst power, channel power and so on. Many traditional power meters have been unable to meet the requirements of digital signal power measurement. Some of the power measurement tasks have been completed by spectrum analyzer.

Spectrum analyzer and power meter can measure radio frequency power, which can be divided into absorption power meter and power meter. The same is the power measurement. Different testing instruments and methods focus on different aspects.

Measurement methods of radiofrequency power include:

· Spectrum analyzer measurement
· Absorption power measurement
· Pass-through power measurement

8.2.2. The Vector Network Analyzer Used in the IoT

Imperfections in the test equipment or the test setup have following features:

• Typically predictable

• Can be easily factored out by a user calibration

• Examples that occur across the frequency range [11]:

Its systematic error includes:

- Output power variations

- Ripples in the VNA receiver's frequency response

- Power loss of RF cables that connect the DUT to the VNA.

Its random error includes:

• Error caused by noise emitted from the test equipment or test setup that varies with time

• Determines the degree of accuracy that can be achieved in your measurement

• Cannot be factored out by a user calibration

• Examples include:

- Trace noise.

Its drift error includes:

• Measurement drift and variances that occur over time in test equipment and test setup after a user calibration • The amount that the test setup drifts over time determines how often your test setup needs to be recalibrated • Examples include: - Temperature changes - Humidity changes - Mechanical movement of the setup.

1) Basic VNA Operation

A VNA contains both a source(that is used to generate a known stimulus signal) and a set of receivers(that are used to determine changes to this stimulus caused by the device-under-test or DUT). This illustration highlights the basic operation of a VNA. In simplification, Fig. (8-1) shows the source coming from Port 1, but most VNAs today are multipath instruments and can provide the stimulus signal to either port.

As is shown in Fig. (8-2), the S-Parameter is defined in the below: Scattering parameters or S-parameters describe the electrical properties and performance of RF electrical components or networks of components when undergoing various steady state electrical signal stimuli. They are complex numbers, having both magnitude and phase, and they are normally familiar measurement parameters, for example, the gain, loss, and reflection coefficient [11].

The Smith chart is a very useful tool used to determine complex impedances and admittances of RF circuits. Most network analyzers can automatically display the Smith chart, plot measured data on it, and provide adjustable markers to show the calculated impedance.

Impedance Smith Chart [11]:

1. The circles touching the right corner are constant-resistance circles.
2. The curves stretching from the right corner to the outer edges of the impedance Smith chart are constant-reactance curves.
3. The center of the circle is the Zo point. In most cases, Zo = 50 ohms. This is also the 20-millisiemens (mS) point.

Admittance Smith Chart [11]

1. The circles in the Smith chart that touch the left corner are constant-

conductance circles.
2. The curves stretching from the left corner of the Smith chart to the outer edges of the admittance Smith chart are constant susceptance curves.

The common S-parameter names are shown in Fig. (**8-3**).

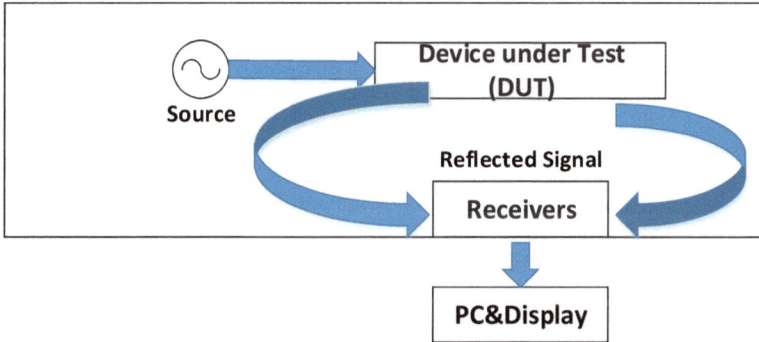

Fig. (8-1). The basic VNA operation.

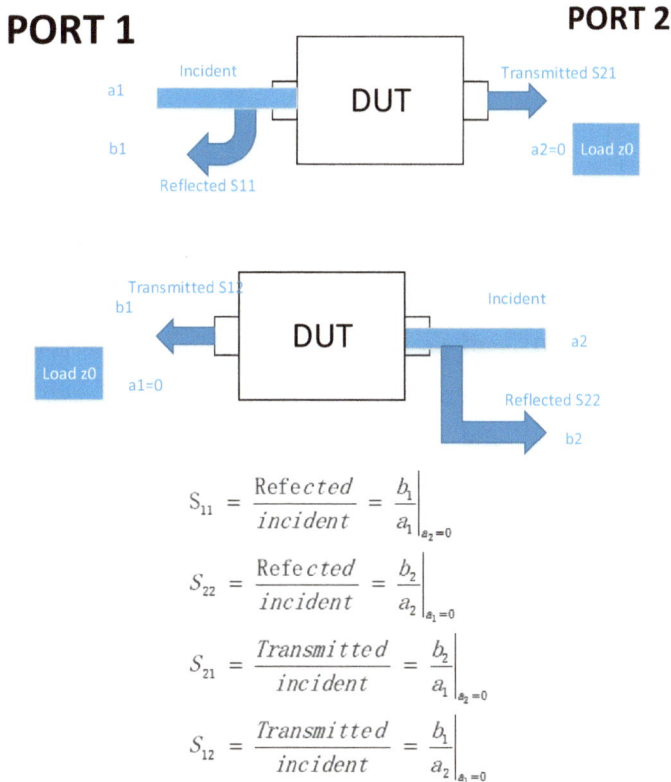

$$S_{11} = \frac{Refected}{incident} = \frac{b_1}{a_1}\Big|_{a_2=0}$$

$$S_{22} = \frac{Refected}{incident} = \frac{b_2}{a_2}\Big|_{a_1=0}$$

$$S_{21} = \frac{Transmitted}{incident} = \frac{b_2}{a_1}\Big|_{a_2=0}$$

$$S_{12} = \frac{Transmitted}{incident} = \frac{b_1}{a_2}\Big|_{a_1=0}$$

Fig. (8-2). The basic S-parameter theory.

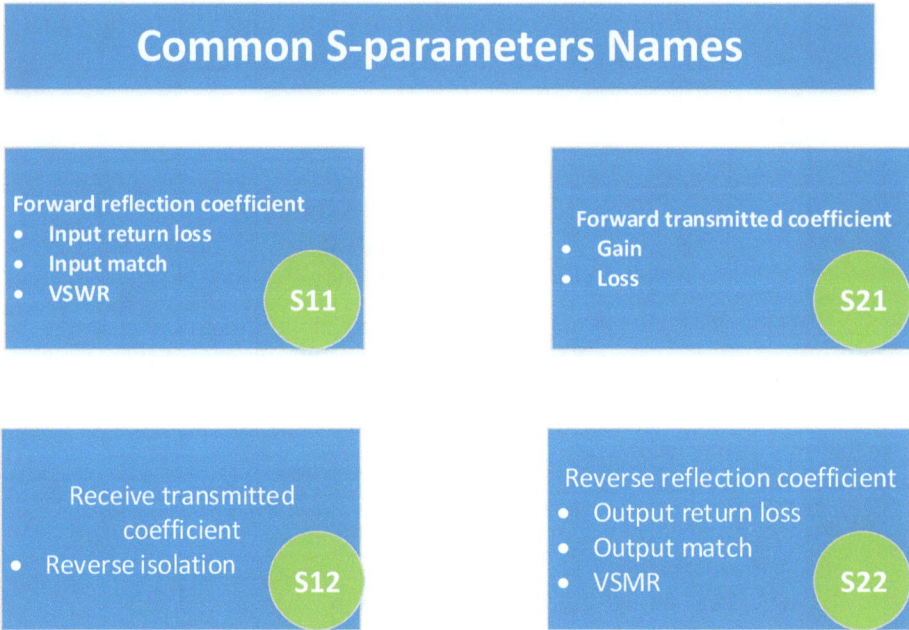

Fig. (8-3). The Common S-parameter names [11].

8.2.3. The Bluetooth Communication Technologies in the IOT [12]

Bluetooth is a wireless technology standard that enables short-range data exchange between fixed devices, mobile devices and building personal area networks (using 2.4-2.485 GHz ISM band UHF radio waves). Bluetooth can connect multiple devices to overcome the problem of data synchronization [4].

Bluetooth technology is a new wireless communication technology jointly announced by Ericsson, Nokia, Toshiba, IBM and Intel in May 1998. Bluetooth device is the main carrier of Bluetooth technology application. Common Bluetooth devices are computers, mobile phones, *etc*. Bluetooth products include Bluetooth modules, Bluetooth radio connections and software applications. The connections between Bluetooth devices must be paired within a certain range. This paired search is called short-range temporary network mode, also known as micro-network, which can accommodate up to eight devices. Bluetooth devices can be connected successfully when there is only one master device and more slaves. Bluetooth technology has radio frequency characteristics. It adopts the structure of TDMA and multi-layer network and applies frequency hopping technology and wireless technology. It has the advantages of high transmission efficiency and high security [12].

1). The underlying hardware module.

The underlying hardware module of Bluetooth technology system is managed by baseband, frequency hopping and link. Baseband is used to transmit Bluetooth data and frequency hopping. The infinite frequency hopping layer is a microwave that does not need the authorization to pass through the 2.4 GHz ISM band. Datastream transmission and filtering are realized in the wireless frequency modulation layer. This paper mainly defines the requirements for Bluetooth transceiver to work normally in this band. Link management implements the security control of link establishment, connection, and removal.

2). Intermediate protocol layer.

The intermediate protocol layer of Bluetooth technology system consists of four parts: service discovery protocol, logical link control, and adaptation protocol, telephone communication protocol and serial simulation protocol. The role of the service discovery protocol layer is to provide a mechanism for upper applications to use services in the network. Logical link control and adaptation protocols are responsible for data disassembly, reuse protocols and quality of service control, and are the basis of other protocol layers.

3). High-level applications.

In Bluetooth technology system, high-level applications are the top part of the framework. The high-level applications of Bluetooth technology mainly include file transfer, network and LAN access. Different kinds of high-level applications are wireless communications implemented by corresponding applications through certain application modes.

8.2.4 The wifi communication technologies in the IOT

"Action Hotspot" is the brand certification of the Wi-Fi alliance manufacturer as a product. It is a wireless LAN technology based on the IEEE 802.11 standard. Based on the close relationship between the two systems, Wi-Fi is often regarded as a synonym of the IEEE 802.11 standard. "Wi-Fi" is often written as "WiFi" or "Wifi", but they are not recognized by the Wi-Fi Alliance.

Not every product that matches IEEE 802.11 applies for Wi-Fi alliance certification. The relative lack of Wi-Fi certification does not necessarily mean incompatibility with Wi-Fi devices.

IEEE 802.11 devices have been installed in many products on the market, such as personal computers, game consoles, MP3 players, smartphones, tablets, printers, laptops and other peripherals that can access the Internet wirelessly [13].

8.3.   THE   STATE-OF-THE-ART   INSTRUMENTS   AND   THEIR APPLICATIONS

8.3.1. The Biopotential Measurements

The ADS1291, ADS1292, and ADS1292R are multichannel, simultaneous sampling, 24-bit, delta sigma ($\Delta\Sigma$) analog-to-digital converters (ADCs) with a built-in programmable gain amplifier (PGA), internal reference, and an onboard oscillator [14].

The hardware of the development kit for ADS1292 includes:

Four electrode ECG cable support ;

View six ECG Leads: Lead I, Lead II, Lead III, Lead a VR, Lead a VL, Lead a VF and respiration channel;

Two modes of operation: Evaluation and Live ECG / Respiration;

Acquire data at up to 8 kHz in Evaluation mode ;

Current based Lead off detection; USB based power and PC application connectivity;

Access to all ADS1x9x registers *via* an easy to use GUI, it has the following features:

• Built-in time domain, histogram, FFT and ECG / Resp related analysis on the PC application ;

• Live ECG with heart rate calculation; Live Respiration wave with respiration rate calculation ;

• USB based firmware upgrade option ;

The MSP430 Firmware debugging can use an ez430 USB emulator [14].

The micro-controller in ADS1292R receives the two-channel data from ADC through the SPI interface to send to the PC. In the middleware, the software is designed to handle the following activities:

• Data acquisition

• ADC Lead off detection

• DC signal removal

• Multiband pass filtering

• ECG lead formation

• QRS (HR) detection

• RR Detection

• USB communication

• Firmware upgrade through USB [14].

8.3.2. The Sensor Tag based on CC2650 and Smartwatch

1) Smartwatch

The modern smartwatches require:

● Compact battery management with ultra-low standby for longer battery run time

● Multi-parameter biosensing to monitor the user's health and physical activity

● Vivid, high-resolution OLED display with capacitive touch and haptic feedback

TDA-00011 is for a complete Optical Heart Rate Monitor wristwatch end equipment featuring TI signal chain, power, and connectivity components. With TI's **AFE4400**AFE, you can accelerate and simplify your wrist-based HRM design while still ensuring measurement performance needed for serious Fitness designs. This reference design also includes a full BLE connectivity design for easy interface to BLE enabled smartphones, tablets, *etc.*

The Heart Rate Monitor (HRM) is an electronic device that detects physiological parameters and converts to usable heart rate reading. Heart rate is the number of times the heart beats in a minute and it is produced *via* depolarization at the sinoatrial and atrioventricular nodes in the heart. A basic HRM is comprised of a sensing probe attached to a patient's earlobe, toe, finger or other body locations, depending upon the sensing method (reflection or transmission), and a data acquisition system for the calculation and eventually display of the heart rate.

2) Sensor tag based on CC2650

The CC2650 µTag is an ultra-compact reference design for the CC26xx family of devices. The PCB measures 141 mm^2, or approximately 16×9 mm^2, and contains the CC2650, a TMP102 temperature sensor, an accelerometer, push button, and an LED. The board also contains a single-ended, discrete component RF filter and a very compact 0402 chip antenna. The board is powered from a CR1620 coin cell (16 mm diameter) battery. There is also a 6-pin Tag Connect interface for cJTAG access on the PCB.

The CC2650 µTag block is shown in Fig. (8-4).

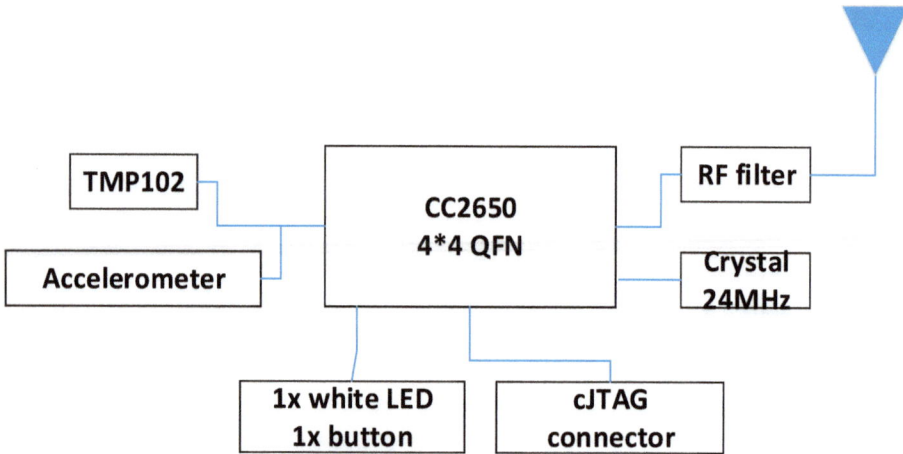

C2650 uTag Black Diagram

Fig. (8-4). The uTag block diagram for CC2650.

The CC2650 device is a wireless MCU targeting Bluetooth Smart, ZigBee™ and 6LoWPAN, and ZigBee RF4CE remote control applications. For more details, refer to the whitepaper, Bringing Wireless Scalability to Intelligent Sensing Applications [15].

The TMP102 device is a digital temperature sensor ideal for NTC and PTC thermistor replacement where high accuracy is required. The device offers an accuracy of ±0.5°C without requiring calibration or external component signal conditioning. IC temperature sensors are highly linear and do not require complex calculations or lookup tables to derive the temperature. The on-chip 12-bit ADC offers resolutions down to 0.0625°C.

The CC2650 μTag is intended both for BLE broadcasting and used as a BLE Peripheral. Thus, the SimpleBLEBroadcaster and SimpleBLEPeripheral are both good starting points for developing the software for the board. If the application requires maintaining BLE connections, the 32 kHz RC oscillator calibration must be enabled since the RTC is derived from this clock source. This process is described in an app note available on the CC2640 product page [15].

PROBLEMS

8-1 Design an intelligent instrument using CC2650.

8-2 Design an intelligent instrument using ADS1292R.

REFERENCES

[1]   D.M. Barakah, and M. Ammaduddin, "A Survey of Challenges and Applications of Wireless Body Area Network (WBAN) and Role of a Virtual Doctor Server in Existing Architecture[C]", *Third International Conference on Intelligent Systems Modelling and Simulation,* 2012 pp. 214-219. [http://dx.doi.org/10.1109/ISMS.2012.108]

[2]   M.J. Buller, W.J. Tharion, S.N. Cheuvront, S.J. Montain, R.W. Kenefick, J. Castellani, W.A. Latzka, W.S. Roberts, M. Richter, O.C. Jenkins, and R.W. Hoyt, "Estimation of human core temperature from sequential heart rate observations", *Physiol. Meas.,* vol. 34, no. 7, pp. 781-798, 2013. [http://dx.doi.org/10.1088/0967-3334/34/7/781] [PMID: 23780514]

[3]   J. Sola, S. Castoldi, and O. Chetelat, "SpO2 Sensor Embedded in a Finger Ring: Desing and implementation", *Proceedings of the 2006 International Conference of the IEEE Engineering In Medicine and Biology Society,* pp. 4495-4498 New York, NY, USA [http://dx.doi.org/10.1109/IEMBS.2006.260820]

[4]   Texas Instruments Incorporated, *LMT70, LMT70A ±0.05°C Precision Analog Temperature Sensor, RTD and Precision NTC Thermistor IC datasheet.* Rev. A, 2015.

[5]   NV Melexis, *MLX90615 Infra-Red Thermometer, datasheet,* 2013.

[6]   A.G. Sensirion, *Datasheet SHT20 Humidity and Temperature Sensor IC,* 2014.

[7]   Maxim Integrated, *MAX30102 High-Sensitivity Pulse Oximeter and Heart-Rate Sensor for Wearable Health, datasheet,* 2015.

[8]   NeuroSky Inc, *NeuroSky BMD101 Product Brief Rev 1.2,datasheet,,* 2012.

[9]   "Microsoft Azure application suite" available from https://en.wikipedia.org/wiki/Microsoft [Accessed: 2019.2.1].

[10]   DMM7510, https://www.tek.com/ [Accessed: 2019.2.1].

[11]   Vector Network Analyzer, https://www.tek.com/ [Accessed: 2019.2.1].

[12]   K.H. Chang, "Bluetooth: A viable solution for IoT? [Industry Perspectives]", *Wireless Communications IEEE,* vol. 21, no. 6, pp. 6-7, 2014. [http://dx.doi.org/10.1109/MWC.2014.7000963]

[13]   Paulo C. Bartolomeu, "Supporting Deterministic Wireless Communications in Industrial IoT", *IEEE Transactions on Industrial Informatics PP.99,* 2018.

[http://dx.doi.org/10.1109/TII.2018.2825998]

[14] Texas Instruments Incorporated, "ADS1292R, 2-Channel 24-Bit ADC With Integrated Respiration Impedance and ECG Front End (Rev. A)",

[15] Texas Instruments Incorporated,CC2650, SimpleLink multi-standard 2.4 GHz ultra-low power wireless MCU ,datasheet (Rev. A).2014.

## CHAPTER 9

The Advance of the Intelligent Instrument Applied in an Online Equipment Monitoring System

Abstract: In this chapter, we will introduce the state-of-the-art instruments of online equipment monitoring system. Meanwhile, the technology of fault tolerance, fault analysis and the fault identify problem will be discussed.

Keywords: Condition monitoring, FTA, Fault identify, Fault tolerance, MindSphere, Movilizer, IIOT, PEMS.

9.1. SOME NEW STATE-OF-THE-ART INSTRUMENTS OR SYSTEMS

9.1.1. State-of-the-Art Instruments of Siemens-MindSphere

Industrial processes in the digital enterprise demand total transparency and a high level of customization. Some solutions close the gap between the real and the digital worlds – and create new potentials to add value along the entire production and supply chain. It already offers the future industry, consistent and end-to-end range of identification and locating systems for customer-specific applications.

Connection to cloud applications maximizes flexibility and ensures that manufacturing will remain viable in the future – MindSphere also allows large volumes of data to be analyzed and used efficiently. Plant availability, capacity utilization, and energy saving potentials are made transparent [1].

SIMATIC RF600 technology(RFID) with built-in OPC UA interoperability opens doors to new applications. OPC-UA can help span wider geographic areas and include cloud-based analytics for predictive capabilities *via* the Siemens MindSphere IoT operating system platform [2].

Production cells and assembly lines inside factories have been digitally automated for decades, keeping human operators informed and automated systems synchronized.

Real-time decision was not supported well. Fortunately, that's changing fast. Limited machine-to-machine communications in the past have given rise to the idea of an Internet of Things (IoT) continuously sharing data with each other.

Supporting the IoT is newer, faster, and more interoperable technologies backed by open communication standards and ubiquitous connectivity. Moreover, emerging cloud technologies have lifted much of the capital and management burdens of procuring and deploying complex hardware and software, making technology solutions less costly, much easier, more capable, and far more scalable.

Advanced analytics, for example, are being used for condition-monitoring and predicting maintenance requirements for equipment operating on offshore oil platforms far at sea but with control rooms building onshore safely.

Siemens SIMATIC RF600 UHF RFID solutions have the recently added benefit of being compliant with the 2016 AutoID Companion Specification to the global OPC Unified Architecture (UA) interoperability communications standards. This vastly expands the potential for UHF RFID technology to extend its reach beyond factory floors to cover entire supply chains. In addition, the RF600 RFID solutions can communicate with the Siemens MindSphere IoT cloud operating system. The latter provides a global platform to manage field level data across wide geographic expanses, while also applying advanced analytics for real-time asset visibility and predictive capabilities to support more informed and faster decision making.

Sanctioned as IEC 62541, it builds upon the time-tested mechanics of the OLE for Process Control (OPC) specification. Its advantages include:

• Open and freely available for unrestricted use without fees;

• Cross-platform, to span programming languages and operating systems, including Linux, Windows, OS X, Android, iOS platforms;

• Service-oriented architecture (SOA), enabling users to model RFID data into an OPC UA namespace;

• Robust security, as defined by IEC/TR 62541-2, to provide for authentication, data integrity, and audit ability and guard against threats, such as eavesdropping, message spoofing and flooding, and others.

Another advantage of incorporating OPC UA into SIMATIC RF600 UHF RFID solutions is that: Field-level data from tools, products, cartons, pallets, and containers can be transmitted to Siemens MindSphere, the highly secure, cloud-based IoT operating system, for real-time analysis and archiving. Analytics, in turn, can support much faster and better-informed decision-making.

MindSphere offers what's called a platform-as-a-service (PaaS), hosted in the global public clouds of Amazon and soon Microsoft Azure, with open interfaces to a growing number of the world's top third-party application developers. Its pay-as-you-go subscription model relieves customers of having to incur the capital expense, operating costs, and management efforts of procuring, deploying, and administering [2].

9.1.2. State-of-the-Art Instruments of Honeywell

To decrease the brand damage and first-time wrong-fix rate, Honeywell provides the Movilizer and Industrial IoT.

1) Movilizer Cloud provides the operation across the company boundaries [3]. It is mostly used in the fields of that enterprises need to implement a mobile strategy; or in the field that if they want to drive their profitability and accelerate their business opportunities.

2) Industrial IoT (IIoT): an end-to-end view of operations and a willingness to think in new ways [4].

(1) Operations will be viewed more holistically

The systems that drive business are becoming increasingly complex — they require greater digitization to realize their potential.

Businesses need to understand how their entire operation works holistically.

(2) IIoT will enable next-tier performance

The first thing a company usually does to be more productive is to go "lean", to trim overhead and inventory, streamline processes, *etc*. But that can only take you so far. The second step is to automate your processes, even if it reaches its limit too.

IoT solutions become the third and biggest evolution companies want to take on. It's putting your data to work. It has fewer limits and tremendous opportunities.

(3) Shiny objects will fade, practical solutions will prevail

If you're looking for technology that's relatively easy to install and delivers a high ROI, technologies like AR and VR. They're light on the impact and create a barrier for customer adoption. The future depends on scalable, lower-cost ways to push performance to produce greater business outcomes.

Being compliant with safety regulations. Once your objectives are clear, it's important to be pragmatic in searching for fresh ideas to solve them.

(4) Data will be the differentiator

Companies tend to spend a lot on analytics, but few get the results they're looking for. Many providers don't have the historical data and insights to apply to their analytics. They don't know the theoretical maximums or how to get there. It takes firsthand knowledge in the lab and real life.

For example, if you want to know the shortest distance between two destinations, that's a math formula. The second type of data is empirical — weather, no-go zones and other restraints in getting from A to B.

The third type is big data. Factors you get from collected data, for example, how ambient temperatures and other factors affect your model. Not a fundamental yield, but a maximum yield. What happens when diskless workers don't follow maintenance workflows exactly.

(5)The process will move from reactive to proactive

The future of IIoT relies on "things" — 50 billion devices by 2022 — becoming more and more connected. And those things will continue to learn. The more you can create a closed loop system; the more collected data can inform and improve performance.

Real-time data, combined with predictive analytics, is helping industrials identify warning signs before they become failures. It's turning workers into experts, able to predict machinery problems without affecting production or causing expensive downtime.

The future of any business is hiding in their data — to provide more accurate models that drive more predictive approaches. As things become more connected and machines continue to learn, enterprises will see stronger business outcomes and reach a point of autonomous problem-solving.

(6) IIoT will spark continued innovation

IIoT has the potential to do more than optimize your business. It can revolutionize it. Just as on-site storage became the cloud, and expensive physical servers turned virtual, IIoT allows you to reimagine what's possible for your business.

(7) Small improvements will have profound results

(8) IIoT disruptors will own the future

Example 9-1: Pressure sensor

Solution 9-1:

In the rugged industrial manufacturing environment, vehicle health and maintenance are important (see Fig. **9-1**). Honeywell FP5000 sensors can help monitor:

- Oil pressure
- Coolant pressure
- Fuel flow, consumption, flow, and pressure
- Intake airflow
- Manifold vacuum testing
- Transmission pressure testing
- Engine and emissions testing

To ensure quality, FP5000 Series transducers can be used in component testing (See Fig. **9-1**) for the following:

• Fluid pump pressure • Brake system testing • Suspension pressure • Hydraulic pressure testing.

Throughout HVAC systems, there are many options for FP5000 sensors to be utilized, including cooling towers, and pumps to monitor system.

In injection/blow molding machines, it's important to monitor for even flow and prevent bursting.

With packaging machines, transducers monitor pressure as the bag/pouch is filled; once a specific pressure is reached, the machine stops either sealing or filling the bag. Through measurement of airline pressure, sensors provide mechanism control during grinding with go/no-go feedback to stop the process if there's a manufacturing problem, thereby, reducing scrap as the sensors stop the process.

Fig. (9-1). Model FP5000 Pressure Transducers on COMPONENT TESTING [5].

Subcontractors test and certify hydraulic systems before shipping to the aerospace manufacturer as part of the acceptance testing for product quality control. Test validation assures that the systems are properly functioning through the collection of data, control data, and validation results. Pressure transducers play an important role in measuring and certifying the flow and rate of the hydraulic fluid within the aircraft actuator being tested, and may provide:

• Design validation such as testing an air wing or other subassembly aspects on aircraft or systems by monitoring various pressure lines

9.1.3. State-of-the-Art Instruments of ABB

ABB delivers digitally enabled measurement and analytics products and solutions; their goal is to make selection and ownership easy.

It is important to acquire the emission of pollutants. And in many countries, it is a standard, legally enforced requirement for the process industry to prevent the factory from releasing the emission that does not exceed the thresholds defined by the regulations.

The ABB uses the technology of Artificial intelligence (AI) to offer some solutions for measuring and recording air pollutant emissions that are recognized and accepted by most of the environmental authorities [6].

1) Predictive emission monitoring systems (PEMS)

PEMS [7] include hardware and software, it is mainly a software-based technology that can estimate pollutant concentration by models. The model includes the fundamental model and empirical model. Modern digital processing algorithms are often used to analyze these models. For example, multilinear regressions (MLR), artificial neural networks (ANN), genetic algorithms [7].

PEMS had been used in many factories, for example fluid catalytic cracking (FCC) and sulfur recovery units (SRUs) of the refinery manufactory.

The normal pollutant components to be tested include:

- SO_2
- CO
- NO
- O_2
- flue gas flowrate
- particulate

9.2. THE FAULT TOLERANCE OF CONDITION MONITORING

Fault Tolerance is tolerance of fault, rather than Error. For example, in a two-machine fault-tolerant system, when one machine is in trouble, another machine can take its place to ensure the normal operation of the system. This is common in the early days when computer hardware was not particularly reliable. Although the hardware is much more stable and reliable than before, hardware fault tolerance is still a very important way for systems that do not allow errors.

Mechanical condition monitoring, inspecting and monitoring the working state of the whole machinery equipment or its parts in operation, to judge whether its operation is normal, whether there are signs of abnormality and deterioration, or to track the abnormal situation, predict its deterioration trend, determine its deterioration and wear degree, *etc*. This method is called machinery. Condition Monitoring. The purpose of mechanical condition monitoring is to grasp the abnormal symptoms and deterioration information before equipment failure, to take pertinent measures in advance to control and prevent the occurrence of failure, to reduce the downtime of failure and the loss caused by downtime, reduce maintenance costs and improve the effective utilization rate of the equipment. To monitor the equipment on-line or on-line in use, it is helpful to determine the actual characteristics of the equipment, to make full use of the potential of the equipment and parts, to avoid excessive maintenance, to save maintenance costs and to reduce downtime losses. Especially for automatic lines, programs, production lines or complex key equipment, the significance is more prominent.

Using all kinds of measurement, analysis and discrimination methods to make clear the running state of the machine, combining with the historical state and operating conditions of the machine, lay a good foundation for the performance evaluation, rational use, safe operation and fault diagnosis of the machine [8].

Equipment fault early warning technology can be divided into three categories: mechanism-based approach, knowledge-based approach and data-driven approach. The method based on the mechanism model is the earliest and most in-depth method of fault early warning and condition monitoring. It mainly includes two stages: (1) Residual generation stage: Estimating system output by establishing an accurate mathematical model based on equipment operation mechanism and comparing it with actual measurement value to obtain residual. This stage is constructed. The model is also called residual generator; (2) Residual evaluation stage: residual analysis is carried out to determine whether there is a fault in the process and to further identify the fault type. This kind of method combines closely with control theory, and mainly uses three kinds of concrete methods, namely parameter estimation, state estimation and equivalent space, to construct a residual sequence. Among them, the state estimation method is the most commonly used one, which can be realized by the observer or Kalman filter [9].

The knowledge-based method is mainly based on the heuristic experience knowledge of relevant experts and operators. It describes qualitatively or quantitatively the connection between units and fault propagation modes in the process. After the abnormal symptoms of equipment occur, the reasoning ability of process experts in monitoring is simulated by reasoning and deduction. Actively complete equipment failure early warning and equipment monitoring. This kind of method does not need a precise mathematical model but has a strong dependence on expert knowledge. Commonly used methods include expert system, fault decision tree, directed graph, fuzzy logic, *etc*.

Based on the data-driven method, the mathematical model and the state of the process are established by mining the intrinsic information in the process data, and the effective monitoring of the process is implemented according to the model. With the wide application of intelligent instrument and computer storage technology, massive process data can be effectively monitored, collected and stored. This kind of method is based on such massive data. In monitoring and early warning algorithms, it can be divided into signal processing, rough set, machine learning, information fusion and multivariate statistics. Among these five kinds of algorithms, machine learning algorithm is the most active branch in theory and practice. It includes Bayesian classifier, neural network, support vector machine, k-nearest neighbor algorithm, clustering algorithm, principal component analysis and other algorithms.

The monitoring method based on mechanism model can combine physical knowledge with monitoring system. The way of fault early warning through residual analysis is more conducive to the understanding of professionals. However, most of the mechanism models are simplified linear systems, so they face the complex of non-linearity, a high degree of freedom and multi-variable coupling. In addition, it may cost a lot to build a mechanism model for complex systems. Moreover, the impact of noise in the actual industrial process and the change of environmental factors all increase the risk of model failure. The above reasons make the monitoring method based on mechanism model ineffective and not widely used.

A knowledge-based monitoring method uses a qualitative model to realize early warning and monitoring. When the monitored object is relatively simple and the process knowledge and production experience are sufficient, its performance is better. However, it should be noted that the early warning accuracy of this kind of method is strongly dependent on the richness of expert knowledge and the level of expert knowledge in the knowledge base; at the same time, it is difficult for some experts to describe their practical experience in a reasonably formal way, and there may be "conflicts" when the system is more complex. In addition, the generality of these methods is poor, and the integrity of prior knowledge is generally difficult to guarantee.

Data-driven fault early warning and condition monitoring technology directly build a fault early warning model through the historical data of the system. It does not need to know the precise mechanism model of the system. Therefore, it has strong versatility and adaptive ability. However, because these methods are not clear about the internal structure and mechanism information of the system, it is relatively difficult to analyze and interpret the early warning results. In addition, data-driven algorithms such as machine learning are mainly applied to fault diagnosis, while data-based early warning technology is still in its infancy, reliable and effective. Moreover, due to a large amount of data and the high time complexity of data-driven algorithms, how to improve the efficiency of monitoring algorithms is also an urgent problem to be solved [10].

In addition, fault-tolerant systems are characterized in terms of both planned service outages and unplanned service outages. These are usually measured at the application level and not just at a hardware level. The figure of merit is called availability and is expressed as a percentage. For example, a five nines system would statistically provide 99.999% availability.

Fault-tolerant systems are typically based on the concept of redundancy.

9.3. THE FAULT ANALYSIS [11]

1) Event symbols

Event symbols include primary events and intermediate events.

Primary events are basic event or element of fault system;

Intermediate events are produced by the primary event, it cannot be analyzed singly, it is the output of a gate.

The event symbols are shown in Table **9-1**:

Table 9-1. Event symbols.

Symbols				
Basic Event	**External Event**	**Undeveloped Event**	**Conditioning Event**	**Intermediate Event**
Basic event - failure or error in a system component or element (example: switch stuck in an open position)	External event - normally expected to occur (not of itself a fault)	Undeveloped event - an event about which insufficient information is available, or which is of no consequence	Conditioning event - conditions that restrict or affect logic gates (example: mode of operation in effect)	An intermediate event gate can be used immediately above a primary event to provide more room to type the event description

2) FTA analyze the results from many events and gates

The logic gate is OR gate (any input can produce the output), AND gate (all input can only produce output), Exclusive OR gate (exactly on input produce an output), Priority AND gate (the inputs all in a specific sequence by a conditioning event can produce output), Inhibit gate (an enabling condition specified by a conditioning event can produce output).

3) Transfer symbols

4) Basic mathematical foundation

Events occur follow certain statistical probabilities. For example, the failure probability rate is λ and the costume time is t, then has Equation 9-1,9-2.

$$P = 1 - \exp(-\lambda t) \qquad (9\text{-}1)$$

$$P \approx \lambda t, \lambda t < 0.1 \qquad (9\text{-}2)$$

The Gates in FTA operate follow the probabilities related to the set operations of the Boolean logic set.

So, the output of the median event is probabilities event, it may occur or may not occur.

The probabilities of AND gate is shown in Equation 9-3, it represents a combination of two independent events.

$$P (A \text{ and } B) = P (A \cap B) = P(A) P(B) \qquad (9\text{-}3)$$

The probabilities of an OR gate is given by Equation 9-4.

$$P (A \text{ or } B) = P (A \cup B) = P(A) + P(B) - P (A \cap B) \qquad (9\text{-}4)$$

In most cases, have Equation 9-5:

$$P (A \text{ or } B) \approx P(A) + P(B), P (A \cap B) \approx 0 \qquad (9\text{-}5)$$

The probabilities of an exclusive OR gate is Equation 9-6:

$$P (A \text{ xor } B) = P(A) + P(B) - 2P (A \cap B) \qquad (9\text{-}6)$$

5) Analysis

A. Defining FTA objectives [11]

Fault tree is used to decompose security requirements as part of PSSA? Or as part of SSA to validate security requirements? Fault tree is used for qualitative evaluation? Quantitative assessment? Or both?

Determining the objectives of FTA will help analysts to determine the scope of work. In the PSSA process, fault tree can be used to allocate quantitative probabilistic requirements. When carrying out FTA, we usually use the "predicted" inefficiency based on engineering experience, and the "predicted value" usually requires higher inefficiency than the mathematical allocation of fault tree. In addition, FTA can also be used to guide system architecture design by examining the qualitative objectives of failure-safety.

In the process of SSA, the fault tree needs to use real failure rate data (usually from FMEA/FMES) to check the compliance of security requirements from bottom to top.

B. Defining the depth of FTA analysis.

At what level does FTA need to go deep into the system? Is it necessary to subdivide the system into multiple layers to support multi-level FTA?

Determining the depth of FTA analysis is helpful for analysts to determine the scope of work.

FTA is carried out at different levels, such as aircraft, system and component. Its working boundary and content are different [11].

C. Define "undesirable events"

When conducting FTA, a list of "undesirable events" should be compiled, that is, "top events" list. Top events can be directed directly to FC in FHA or to events in other fault trees (if the fault tree is subdivided into multiple levels).

Different levels of FTA have different sources of top events.

D. Analysis of the Failure and Combination of Top Events

Analysts should collect complete system design data, including the architecture and description of the system and the interconnected system, and make a comprehensive analysis of these data to determine the failure events and their combinations that may cause top events.

In the analysis process, reasonable assumptions should be made for the situation that is difficult to analyze and need to be simplified. On the basis of no impact on the results, the analysis process should be simplified or the analysis process can be carried out smoothly.

E. Building Fault Tree

A clear and accurate description of top events should be adopted to clarify the requirements of the failure rate of top events.

Direct, sufficient and minimal intermediate events should be used to expand the upper and middle levels of the fault tree step by step, and the corresponding logical gates should be used to connect the events.

Expand each intermediate event down until the source of the fault or no further expansion is required.

Allocate predicted probability indicators to assess whether security requirements can be met (PSSA process). Or use real failure rate to verify that security requirements have been met (SSA process).

• Analysis and summary of FTA results

Qualitative and quantitative analysis of FTA results can be carried out after the construction of fault tree.

F1. Fault Tree Qualitative Analysis

The minimum cut set of fault tree represents a failure mode that causes top events. It can be used to qualitatively evaluate the importance of failure events and to conduct common cause analysis.

The definitions of Cut Sets and Minimum Cut Sets are introduced here.

Cut set: A set of several bottom events of a single fault tree that occur and cause the top event to occur.

Minimum cut set: A cut set in which the number of bottom events can no longer be reduced. That is, after any bottom event is removed from the minimum cut set, the remaining set of bottom events is not a cut set. Independence should be guaranteed between events in the minimal cut set.

Nowadays, with the support of fault tree software, we can generally derive the minimum cut set without manual logic operation. However, analysts still need to verify the independence of all "and door" events, which is an important part of the common cause analysis.

Also, as part of the PSSA process, the minimum cut set of fault tree can be used to allocate the functions defined by ARP4754A and the project development assurance level (FDAL/IDAL).

F2. Fault Tree Quantitative Analysis

After determining the minimum cut set of the fault tree, it is necessary to determine the failure rate, exposure time or hidden fault check interval and sequence factor of all bottom events. Finally, the numerical calculation of the fault tree is carried out. If the calculation results of top events can not meet the requirements of the indicators, measures such as design modification and conservative evaluation should be taken.

After completing the qualitative and quantitative analysis of fault tree, the report of fault tree analysis should be compiled.

9.4. THE FAULT IDENTIFIES AND DIAGNOSIS PROBLEM [12]

The main tasks of fault diagnosis are fault detection, fault type judgment, fault location, and fault recovery. Among them: fault detection refers to sending detection signals periodically to the lower computer after establishing a connection with the system, and judging whether the system has a fault by receiving the response data frames; fault type judgment refers to the type of system fault after the system has detected the fault, through analyzing the reasons; fault location is before, On the basis of the two parts, the types of faults are refined, the specific fault locations and causes are diagnosed, to prepare for fault recovery. Fault recovery is the last and most important link in the whole process of fault diagnosis, and different measures should be taken to recover the system faults according to the causes of the faults.

(1) Diagnosis Method Based on Expert System

The diagnosis method based on an expert system is one of the most noticeable development directions in the field of fault diagnosis. It is also a kind of intelligent diagnosis technology with the most research and application. It has roughly gone through two stages of development: fault diagnosis system based on experience knowledge of experts in the shallow knowledge field and fault diagnosis system based on model knowledge of deep knowledge diagnosis object.

1) Intelligent Expert Diagnosis Method Based on Shallow Knowledge

Shallow knowledge refers to the empirical knowledge of domain experts. Fault diagnosis system based on shallow knowledge obtains diagnosis results by deductive reasoning or production reasoning. Its purpose is to find a fault set so that it can give a given fault.

The reasons for the occurrence of collections are best explained by the presence and absence of symptoms.

Fault diagnosis method based on shallow knowledge has the advantages of direct knowledge expression, unified form, high modularity and fast reasoning speed. But there are also limitations, such as incomplete knowledge set, easy system for problems not taken into account. In a dilemma; weak interpretation ability of diagnostic results and other shortcomings.

2) Intelligent Expert Diagnosis Method Based on Deep Knowledge

Deep knowledge refers to knowledge about the structure, performance and function of diagnostic objects. Fault diagnosis system based on deep knowledge requires that each environment of the diagnostic object has a distinct input-output relationship. First, the inconsistency between the actual output and the expected output of the diagnostic object is used to generate a set of reasons for the inconsistency. Then, according to the domain of the first law knowledge, a set of reasons for the inconsistency is generated. And its knowledge with a clear scientific basis, its internal specific constraints, using a certain algorithm to identify possible sources of failure. An intelligent expert diagnosis method based on deep knowledge has the advantages of convenient knowledge acquisition, simple maintenance and good completeness, but its disadvantages are large search space and slow reasoning speed.

3) Intelligent Expert Hybrid Diagnosis Method Based on Shallow Knowledge and Deep Knowledge

As far as complex equipment systems are concerned, it is difficult to properly accomplish diagnostic tasks whether using shallow knowledge or deep knowledge alone. Only by combining the two, can the performance of diagnostic systems be optimized. Therefore, to make fault intelligent diagnosis systems possess knowledge similar to human expert's ability when building an intelligent diagnosis system, more and more researchers emphasize not only the experience knowledge of experts in the field but also the knowledge of structure, function and principle of diagnosis object. The focus of research is shallow knowledge and deep knowledge. Integrative representation method and usage method. A high-level domain expert always combines his deep knowledge and shallow knowledge to complete the diagnostic task when solving diagnostic problems. Generally, shallow knowledge is preferred to find the solution or approximate solution of the diagnosis problem, and deep knowledge is used to obtain the exact solution of the diagnosis problem when necessary.

(2) Artificial Intelligence Diagnosis Method Based on Neural Network

In knowledge acquisition, the knowledge of the neural network does not need to be collated, summarized and digested by knowledge engineers, but only needs to be trained by examples or examples of domain experts solving problems. In knowledge representation, the neural network adopts implicit representation and expresses some knowledge of a certain problem. In the same network, it has high versatility and is easy to achieve knowledge acquisition and parallel associative reasoning. In the aspect of knowledge reasoning, neural network realizes reasoning through the interaction of neurons.

Previous fault diagnosis systems have been applied in many fields, such as chemical equipment, nuclear reactor, steam turbine, rotating machinery, and motor, and achieved good results. Because the knowledge learned from fault cases by neural networks is only some distributed weights, not the production rules of logical thinking of experts in similar fields, the process of diagnostic reasoning can not be explained and lacks transparency.

(3) Artificial Intelligence Diagnosis Method Based on Fuzzy Mathematics

The fault state of many diagnostic objects is fuzzy. An effective way to diagnose such faults is to apply the theory of fuzzy mathematics. The diagnosis method based on fuzzy mathematics does not need to establish an accurate mathematical model (membership function). The intelligent fuzzy diagnosis can be realized by using local functions and fuzzy rules appropriately and making fuzzy reasoning.

(4) Artificial Intelligence Diagnosis Method Based on Fault Tree

The fault tree method is a searching process that the computer automatically generates the fault tree based on the prior knowledge of the fault and its causes and the knowledge of the fault rate. The diagnostic process begins with a fault of the system "why does it appear". A step-by-step fault tree is constructed along the fault tree, and the root cause of the fault is finally found through a heuristic search of the fault tree. In the process of questioning, the effective and reasonable use of timely and dynamic data of the system will help to carry out the diagnosis process. Fault tree diagnosis method, similar to human thinking mode, is easy to understand. It is widely used in practice but mostly used in combination with other methods.

PROBLEMS

9-1 Read the article References [15] and calculate the measuring vibration severity using simulate data.

9-2 Brief design the reliability of an online vibration monitoring WSN system.

9-3 Design a WSN vibration detection node based on Arduino.

REFERENCES

[1]   Q. Wang, "Siemens launches MindSphere Open Industrial Cloud", *Light Industry Machinery,* vol. 3, pp. 48-48, 2016.

[2]   "This is MindSphere", https://new.siemens.com/global/en/products/software/mindsphere.html

[3]   "The Cloud For Field Operations - Movilizer", https://new.siemens.com/global/en/products/ software/mindsphere.html

[4]   Stan Schneider, "The industrial internet of things (IIoT)", *Internet of Things and Data Analytics Handbook.,* 2017.

[5]   Honeywell Co, "FP5000 application nots",

[6]   S.M. Zain, and K.K. Chua, "Development of a neural network Predictive Emission Monitoring System for flue gas measurement", *IEEE International Colloquium on Signal Processing & Its Applications,* 2011
        [http://dx.doi.org/10.1109/CSPA.2011.5759894]

[7]   T.W. Chien, H.T. Hsueh, H. Chu, W.C. Hsu, Y.Y. Tu, H.S. Tsai, and K.Y. Chen, "A feasibility study of a predictive emissions monitoring system applied to taipower's nanpu and hsinta power plants", *J. Air Waste Manag. Assoc.,* vol. 60, no. 8, pp. 907-913, 2010.
        [http://dx.doi.org/10.3155/1047-3289.60.8.907] [PMID: 20842930]

[8]   S. Nandi, H.A. Toliyat, and X. Li, "Condition monitoring and fault diagnosis of electrical motors-a review", *IEEE Trans. Energ. Convers.,* vol. 20, no. 4, pp. 719-729, 2005.
        [http://dx.doi.org/10.1109/TEC.2005.847955]

[9]   F.B. Schneider, "Implementing fault-tolerant services using the state machine approach: a tutorial", *ACM Comput. Surv.,* vol. 22, no. 4, pp. 299-319, 1990.
        [http://dx.doi.org/10.1145/98163.98167]

[10]   Michel Kinnaert, "Diagnosis and Fault-Tolerant Control",

[11]   Nasser Tleis, *Power Systems Modelling and Fault Analysis.,* 2014.

[12]   T. Marwala, and H.E.M. Hunt, "FAULT IDENTIFICATION USING FINITE ELEMENT MODELS AND NEURAL NETWORKS", *Mech. Syst. Signal Process.,* vol. 13, no. 3, pp. 475-490, 1999.
        [http://dx.doi.org/10.1006/mssp.1998.1218]

[13]   Huai Wang, Ma K, and Blaabjerg F, "Design for reliability of power electronic systems", *Conference of the IEEE Industrial Electronics Society,* 2012

[14]   E Bauer, and R. Adams, "Design for Reliability of Virtualized Applications",
        [http://dx.doi.org/10.1002/9781118393994.ch12]

[15]   X. Xiaoyan, "Design of Wireless Sensor Networks Node Applied to Acquisition and Transmission of Vibration Signals", *International Conference on Computational Aspects of Social Networks,* 2010
        [http://dx.doi.org/10.1109/CASoN.2010.158]

CHAPTER 10

The Application of Intelligent Instrument: A Mobile Robot Instrument

Abstract: In this chapter, the concept of the moving robot and the intelligent instrument will be introduced, followed by the smart small car and the intelligent instrument, and the micro UAV and the intelligent instrument. In the end the state-of-the-art instruments and their applications will be presented.

Keywords: Attitude sensors, CCD, Moving robot, PID control, Smart car, UAV.

10.1. THE CONCEPT OF THE MOVING ROBOT AND THE INTELLIGENT INSTRUMENT

A robot is an automation machine that can do some complex work like people, and so on [1].

A robot is a machine that automatically executes work. It can not only accept human command but also run pre-programmed procedures. It can also act according to the principles and programs formulated by artificial intelligence technology. Its task is to assist or replace human activity, such as production, construction, or dangerous work.

Robots are generally composed of actuators, actuators, detection devices, control systems and complex machinery.

In the robot body, its arm generally adopts a spatial open-chain linkage mechanism, in which the motion pair (rotating pair or moving pair) is often called joints, and the number of joints is usually the degree of freedom of the robot. According to the different types of joint configuration and motion coordinates, robot actuators can be divided into rectangular coordinates, cylindrical coordinates, polar coordinates and joint coordinates. For anthropomorphic considerations, the relevant parts of the robot body are often referred to as the base, waist, arm, wrist, hand (gripper or end effector) and walking part (for mobile robots).

Driving device: It is the mechanism that drives the actuator to move. According to the command signal of the control system, the robot can move through the power element. It inputs electrical signals and outputs lines and angular displacements. The main driving devices used by the robot are electric driving devices, such as stepping motors, servo motors, *etc*. In addition, hydraulic and pneumatic driving devices are also used.

Detection device: Its tasks include detecting the robot's movement and working conditions; to feedback to the control system according to needs; to adjust the executing mechanism to ensure that the robot's action meets the predetermined requirements after comparing it with the set information. Sensors as detection devices can be roughly divided into two categories: one is internal information sensors, which are used to detect the internal status of each part of the robot, such as the position, speed, acceleration of each joint, and send the measured information as a feedback signal to the controller to form a closed-loop control. The other is the external information sensor, which is used to acquire information about the robot's working object and external environment, so that the robot's action can adapt to the changes in external conditions, so that it can achieve a higher level of automation, and even make the robot have some kind of "feeling" and develop to intellectualization, such as vision, acoustics and so on. External sensors give information about the working objects and working environment and use this information to form a large feedback loop, which will greatly improve the working accuracy of the robot.

Control system: One is the centralized control, that is, the whole control of the robot is carried out by a microcomputer. The other one is decentralized control; *i.e.* multi-computers are used to share the control of the robot. When the upper and lower computers control the robot together, the host computer manages the system, communicate, calculate kinematics and dynamics, and send command information to the lower computer. Each joint corresponds to a CPU and performs interpolation and servo control to achieve the given motion and feedback information to the host. According to different task requirements, the control mode of the robot can be divided into point control, continuous trajectory control and force (moment) control.

Robots can be divided into two categories: industrial robots and special robots. The so-called industrial robots are multi-joint manipulators or multi-degree-of-freedom robots oriented to the industrial field. Besides industrial robots, special robots are all kinds of advanced robots used in non-manufacturing industries and serving human beings, including service robots, underwater robots, entertainment robots, military robots, agricultural robots, robotic machines and so on. In special robots, some branches have developed rapidly, and there is a trend of independent system, such as service robots, underwater robots, military robots, micro-manipulation robots and so on. International robotics scholars divide robots into two categories: industrial robots in manufacturing environments and service robots in non-manufacturing environments, which are consistent with the classification of humanoid robots in China.

10.2. THE SMART SMALL CAR AND THE INTELLIGENT INSTRUMENT

An example of a smart small car is discussed in the following:

The Intelligent Vehicle System uses Freescale's 32-bit microcontroller MK60DN512ZVLQ10 as the core control unit for the control of the Intelligent Vehicle System [2].

The optoelectronic sensor scheme is adopted in the intelligent vehicle system, the position signal of the racing car is received by linear CCD installed on the rocker steering gear through the AD port of K60 MCU, which is used for the decision-making of the racing car's motion control.

At the same time, the PWM wave is emitted inside, the DC motor is driven to control the acceleration and deceleration of the intelligent car, and the servo steering gear pair is used. The steering control of the racing car enables the racing car to follow its right track [3].

To control the speed of the car accurately, a photoelectric encoder is installed on the output shaft of the motor of the intelligent car. The pulse signal of the encoder when it rotates is collected, and the digital PID closed-loop control is carried out regularly after being captured by K60.

Besides, LCD keyboard and dial switch are added as input and output devices for speed and control strategy selection of intelligent vehicles.

The block diagram of the overall structure of the system is shown in Fig. (**10-1**).

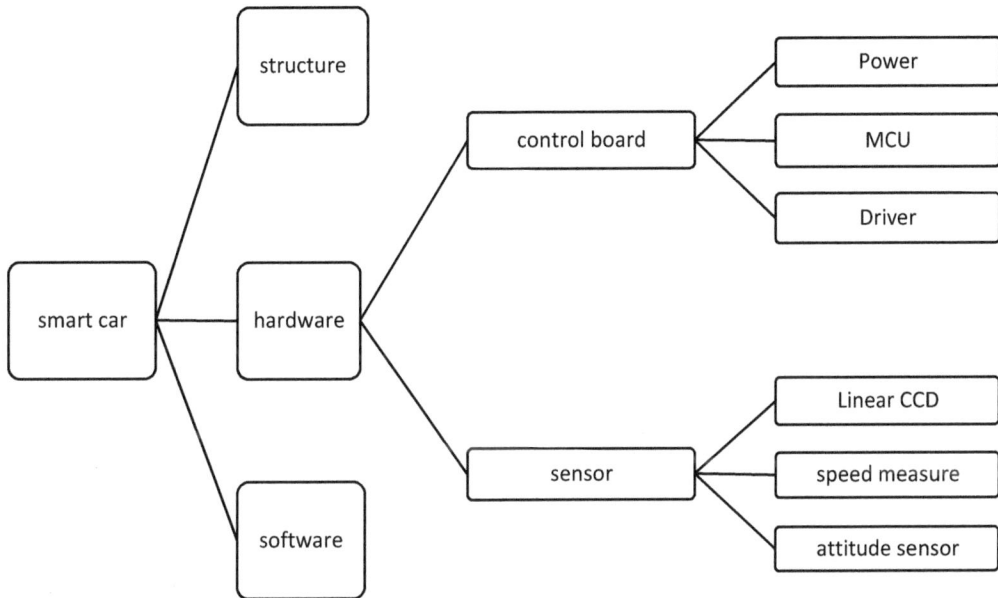

Fig. (10-1). The block diagram of a small smart car.

According to the above system design, the car consists of six modules: MK60DN512ZVLQ10 main control module, sensor module, power module, motor drive module, speed detection module and auxiliary debugging module. The functions of each module are as follows:

The main control module of MK60DN512ZVLQ10 collects the signals of photoelectric sensors, photoelectric encoders and other sensors, controls decisions according to the control algorithm, and drives DC motor and servo motor to complete the control of intelligent vehicles.

Linear CCD senses the track information in front of it and provides the necessary basis and sufficient reaction time for the intelligent vehicle's main control module to make a decision.

The power module provides a stable power supply for the system.

The motor drive module controls the acceleration and deceleration of the motor.

Speed detection module, which detects the speed of intelligent vehicle wheels, is used for speed closed-loop control.

Attitude sensor is mainly used for judging the uphill of intelligent vehicles.

Fig. (10-2). The profile of a small smart car [4].

The model car contacts with the ground through four tires and the coaxial of two rear wheels is limited and cannot be adjusted, which keeps parallel with the forward direction of the model car. Therefore, to change the contact mode between the model car and the ground, adjust the four-wheel positioning which is beneficial to the steering and straight line of the model car, it can only be achieved by adjusting the positioning parameters of the front wheel.

The front-wheel parameters of model B car can be adjusted are back inclination angle, main inclination angle and wheel front harness. The three parameters can be adjusted.

The so-called main inclination is to incline the upper end of the board(*i.e.* the steering axis) inward. Looking from the front of the car, there is an angle between the main pin axis and the vertical line passing through the center of the front wheel, that is, the main pin inclination angle. The function of the main inclination is to make the wheels automatically correct and turn lightly in time after steering.

The so-called backward tilt is to tilt the upper end of the board (*i.e.* the steering axis) slightly backward. Looking from the side of the car, there is an angle between the main pin axis and the vertical line passing through the center of the front wheel, that is, the rear inclination angle of the main pin. The function of the mainboard rollback is to increase the stability of the vehicle when running in a straight line and to make the front wheel return automatically after steering.

From the front of the vehicle, the distance between the front end and the rear end of the left and right tires is measured at the height of the two axles. The difference between the front-end distance and the rear end distance is the front bundle value. The distance between the front end and the back end is the negative front bundle, and vice versa, the positive front bundle. Relevantly, there are front bundle angles, *i.e.* the angle at which the wheels on both sides of the horizontal plane incline inward from the front.

The height of the gravity center obviously affects the stability of the intelligent vehicle. When the center of gravity is on the high side, the smart car is likely to roll over during the telling turn.

The adjustment of the entity's center of gravity mainly proceeds from the following two aspects:

First, height adjustment of chassis: the lower the center of gravity of the intelligent vehicle, the better, the more direct way to achieve the center of gravity drop when reducing the site. However, due to the existence of ramps in the track, too low chassis may not be conducive to the uphill.

Secondly, height adjustment of car body components: in the process of intelligent vehicle refitting, we have always considered the center of gravity as one of the factors. Make the weight distribution as close to the chassis as possible. Besides, smaller circuit boards can be embedded in the voids of other components of the chassis.

There are two kinds of commonly used motor drivers: one is the integrated motor driver chip; the other is the design of N-channel MOSFET and special gate driver chip. Among the common integrated H-bridge motor driver chips on the market, 7960 is a very common driver chip in competitions and is a half-bridge driver. Therefore, two chips are needed to achieve positive and negative control. The driving capacity of the single chip is 43A and 7970 is 68A.

Two half-bridge intelligent power driver chips BTS7960B are combined to form a full-bridge driver to drive the DC motor to rotate. BTS7960B is a high-current half-bridge integrated chip for motor drive. It has a high-side MOSFET with a P-channel, a low-side MOSFET with an N-channel and a driving IC. P-channel high-side switch eliminates the need for a charge pump, thus reducing electromagnetic interference (EMI). The integrated driving IC has the functions of logic level input, current diagnosis, slope adjustment, dead time generation, and over-temperature, over-voltage, under-voltage, over-current, stop-up and short-circuit protection.

The typical on-state resistance of BTS7960B is 16m, and the driving current can reach 43A. Adjusting the external resistance of SR pin can adjust the turn-on and turn-off time of MOS transistor and has the function of anti-electromagnetic interference. The IS pin is the current detection output pin. The INH pin is the enabling pin, and the IN pin is used to determine which MOSFET is on. When IN = 1 and INH = 1, the high-side MOSFET turns on and outputs a high level. When IN = 0 and INH = 1, the low-side MOSFET turns on and outputs low level. By controlling the switch action of BTS7960B with the PWM signal of 20 kHz, the forward and backward PWM driving, reverse connection braking and energy consumption braking of the motor are realized.

Because the discrete N-channel MOSFET has very low on-resistance, the total resistance of the armature circuit is greatly reduced. Besides, the specially designed gate drive circuit can improve the switching speed of MOSFET, improve the modulation frequency of PWM control mode, and reduce armature current ripple. And the special gate driver chips usually have the functions of preventing the same arm from turning on, hardware dead-time, under-voltage protection and so on, which can improve the reliability of the circuit.

However, because the internal resistance of the motor is only a few milliohms, the use of integrated chips often results in large heat, which reduces the efficiency of the motor and the stability of speed control.

So, we use discrete N-channel MOSFET with very low on-resistance, which greatly reduces the total resistance of the armature circuit. In addition, the specially designed gate drive circuit can improve the switching speed of MOSFET, improve the modulation frequency of PWM control mode, and reduce armature current ripple. And the special gate driver chips usually have the functions of preventing the same arm conduction, hardware dead-time, under-voltage protection and so on, which can improve the reliability of the circuit.

Using infrared photoelectric to detect the starting line of the tube, the principle of photoelectric sensor to detect road information is that the transmitter transmits a certain wavelength of infrared ray and reflects it to the receiving tube through the ground. Because of the different reflection coefficients on black and white, most of the light is absorbed on black, and most of the light can be reflected back on white, so the received reflected light intensity is different, which makes the collector voltage of phototransistor change. After comparison, the black-an--white information of the track can be obtained.

In an intelligent vehicle system, the linear CCD sensor is very important for path recognition in the control system. MK60DN512ZVLQ10 MCU processes the track information collected by linear CCD sensor, and then controls the intelligent vehicle to drive along the black line.

It is stipulated in the competition that the vehicle models participating in the photoelectric balance group can use photoelectric sensors and linear CCD sensors of the specified type for road detection. Laser sensors are prohibited. If linear CCD is used in photoelectric balance group, the TSL1401 series linear CCD of Texas Advanced Optoelectronic Solution Company is needed. TSL1401 linear CCD sensor consists of 128 linear photodiodes. Each photodiode has its own integration circuit, which we call pixels. The gray value of each pixel is proportional to its perceived light intensity and integration time.

The incremental PID control is used for the shaking head steering gear control, the position PD control is used for the steering gear control, and the position PD control is used for the speed closed-loop control. In this scheme, the trial and error method is used to determine the proportional, integral and differential parameters of the controller.

The tuning of the PID controller parameter is the core content of the control system design. It determines the proportional coefficient, integral time and differential time of the PID controller according to the characteristics of the controlled process. There are many methods to tune the parameters of PID controllers.

Generally speaking, there are two kinds. One is the theoretical calculation tuning method. Based on the mathematical model of the system, the parameters of the controller are determined by theoretical calculation. The calculated data obtained by this method may not be directly used, but must be adjusted and modified through the actual engineering. The other is the engineering tuning method, which mainly depends on engineering experience, it is directly carried out in the test of control system, and the method is simple and easy to use. It is widely used in engineering practice. The engineering tuning methods of PID controller parameters mainly include critical proportion method, response curve method and attenuation method.

The two methods have their own characteristics, and their common points are that the parameters of the controller are tuned according to the empirical formula of engineering. However, no matter which method is adopted, the parameters of the controller need to be adjusted and perfected in the actual operation. Now the critical proportion method is generally used.

Using this method, the parameters of the PID controller are tuned as follows:

(1) Firstly, a sufficiently short sampling period is pre-selected to make the system work.

(2) Only proportional control is added until the critical oscillation of the step response of the system to the input occurs. The proportional amplification factor and the critical oscillation period are recorded.

(3) The parameters of the PID controller are obtained by formula calculation under certain control degree.

If the static error is eliminated by using proportional-integral control, and the dynamic process is still unsatisfactory after repeated adjustments, a differential link can be added to form a proportional, integral and differential controller.

When the pre-differential time is set as zero and it is increased based on the second step. The proportional coefficient and differential time are also changed accordingly, and the satisfactory adjustment effect and control parameters are obtained by trial and error step by step.

10.3. THE SMART UAV AND THE INTELLIGENT INSTRUMENT

Four-rotor aircraft is a kind of unmanned aerial vehicle. It has four propellers that are evenly distributed and cross-symmetrical on the plane and can take off and land vertically [5].

Compared with conventional rotorcraft helicopter, it has a stronger relative load capacity. It has simple structure, strong maneuverability, and maneuverability, and can fly to the area close to the ground target, so it is easy for the operator to control and reconnaissance.

So, it has good development space and practical performance.

Four-rotor UAV controls the rotational speed of four rotors to achieve various flight actions. The lift of the Four-rotor UAV is obtained by four rotors installed at the top of the frame.

Attitude control of Four-rotor UAV is the most basic and key part of the whole control system. As an inner-loop control, it requires a very high accuracy of real-time attitude information and acquisition speed. Attitude angle and attitude angular velocity are acquired by accelerometer and gyroscope respectively, but both sensors have their own shortcomings.

Accelerometer can measure an accurate attitude angle in a static state, but it is very sensitive to mechanical vibration and additional disturbance acceleration. Therefore, it is necessary to fusion with the acquired attitude angle and attitude angular velocity.

Kalman filter algorithm based on optimal estimation has high precision, obvious denoising effect. The arithmetic of average filtering and adaptive filtering is easier to realize and faster to solve, especially for attitude calculation of UAV [6].

PROBLEMS

10-1 Design a moving robot based on Arduino.

10-2 Design a small smart car based on Arduino.

10-3 Design a UAV based on Arduino.

REFERENCES

[1]   Robot, Available online: https://en.wikipedia.org/wiki/Robot

[2]   Z. Ji, I. Ganchev, M. O'Droma, L. Zhao, and X. Zhang, "A cloud-based car parking middleware for IoT-based smart cities: design and implementation", *Sensors (Basel),* vol. 14, no. 12, pp. 22372-22393, 2014.
[http://dx.doi.org/10.3390/s141222372] [PMID: 25429416]

[3]   M. Wang, and Z. Jiang, "The design and comparison of DC Motor Drive Circuit in Smart Car Competition", *IEEE International Conference on Intelligent Computing & Intelligent Systems,* 2009

[4]   Camera team 1 of Beijing University of science and technology, Technical report of smart car competition of China, 2013.

[5]   Nex, Francesco, and F. Remondino. "UAV for 3D mapping applications: a review." Applied Geomatics 6.1(2014):1-15.

[6]   Choi, Seokrim, J. Lim, and S. Yoon. "Forward-Backward Time-Varying Forgetting Factor Kalman Filter Based Landing Angle Estimation Algorithm for UAV (Unmanned Aerial Vehicle) Auto landing Modelling." Lecture Notes in Computer Science 3398(2004):107.

SUBJECT INDEX

A

Ability 15, 71, 72, 84, 109, 110, 149, 195, 212, 232, 250, 251, 257
  adaptive 251
  analytical learning 72
  anti-interference 15
  comprehensive perception 232
  human expert's 257
  reasoning 250
  weak interpretation 257
Absorption power 233
  measurement 233
  meter 233
Absorptivity 228, 229
  molar 229
Activity 2, 104, 121, 122, 238, 239
  internal 121
  physical 239
Adaptive 38
  filter and Noise Canceler 38
  filtering algorithms 38
  spectral line enhancement 38
Advanced analytics 244
  applying 244
Aging 113, 148
  homogeneous 148
Aircraft actuator 248
Air wing 248
Airworthiness statistical notes 109
Algebra calculation principle 114
Algebraic object table 3
Ambient light cancellation (ALC) 227
American aviation commission 109
Amplifier 12, 19, 26, 27, 28, 32, 33, 35, 90, 238
  best phase-locked 33
  buffer 12
  ideal lower noise 27
  low-noise high-frequency 19
  phase-locked 33
  programmable gain 90, 238

Analog 13, 14, 15, 16, 28, 29, 49, 50, 63, 96, 142, 163, 164, 165, 238
  circuit structure 63
  delta modulator 28
  sigma-delta modulator quantizes 29
  to-digital converter (ADC) 13, 14, 15, 16, 28, 49, 50, 96, 142, 163, 164, 165, 238
Analog signal 12, 29, 205
  band-limited 29
  instantaneous input 12
Analysis method 72, 120, 123, 205
  extreme value 123
  inductive 120
  pattern 72
Analysts 254, 255
Analytical thinking parts 7
Analytics 243, 245, 246
  cloud-based 243
  predictive 246
Analytics products 248
Arithmetic average filtering method 57
Artificial 75, 77, 249
  nerve element 75
  neural models 75
  neural network (ANN) 75, 77, 249
Artificial intelligence 2, 71, 72, 248, 258, 260
  diagnosis method based 258
  technology 248, 260
Asset 218, 244
  management decisions 218
  visibility, real-time 244

B

Biological evolution theory 81
Biosignal detection and processing 230
Black-box testing methods 105, 106
BLE 239, 241
  broadcasting 241
  connectivity design for easy interface 239
Blood oxygen saturation 223
Bluetooth 236, 237
  communication technologies 236